'*Rootbound* is both relevant and important, questioning what it means to call oneself a gardener, and where horticulture fits within the modern urban experience'

Gardens Illustrated

'A fascinating insight into the outlook of the millennial generation, who have grown up online, with their private lives made public on Instagram and everything seeming too competitive and fast. The certainty and unhurriedness of the processes of plants seems all the more grounding and necessary by contrast. I loved this book'

Juliet Blaxland

'A beautiful, lyrical story of personal healing that deftly interweaves botanical history and lore and highlights the importance of urban green spaces in an increasingly disconnected world'

Kayte Nunn

'Interweaves gardening history with memoir . . . Vincent's perceptive nature writing and knowledge of horticulture might inspire readers to start planting'

Prospect

Rootbound

Rewilding a Life

Alice Vincent

CANONGATE

This paperback edition published in 2020 by Canongate Books

First published in Great Britain in 2020 by Canongate Books Ltd,
14 High Street, Edinburgh EH1 1TE

canongate.co.uk

1

Some names and identifying details have been changed to protect the privacy of individuals

British Library Cataloguing-in-Publication Data
A catalogue record for this book is available on request from the British Library

ISBN 978 1 78689 772 5

Typeset in Bembo Std by Palimpsest Book Production Ltd,
Falkirk, Stirlingshire

Printed and bound in Great Britain by Clays Ltd, Elcograf S.p.A.

For those who put soil and seeds into my hands

SELECTED PLANT GLOSSARY

Icelandic poppy
Papaver nudicaule

Bracken
Pteridium aquilinum

Sweet pea
Lathyrus odoratus

Buddleja
Buddleja davidii

Swiss cheese plants
Monstera deliciosa

Rosebay willowherb / Fireweed
Chamaenerion angustifolium

Pass-it-on Plant / Chinese money plant
Pilea peperomioides

Cherry blossom
Prunus x yedoensis 'Somei-yoshino'

Auricula
Primula x pubescens

Nasturtium
Tropaeolum majus

Fiddle-leaf fig
Ficus lyrata

Anemone
Anemone coronaria

Purple / false shamrock
Oxalis triangularis

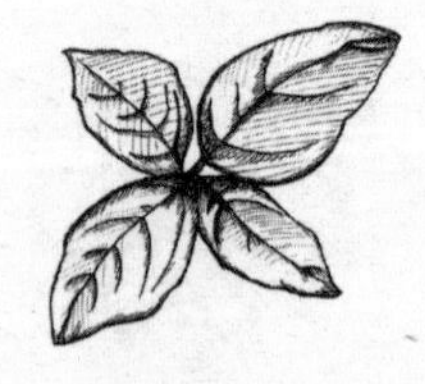

Basil
Ocimum basilicum

CONTENTS

INTRODUCTION

If you got close enough to the metal, you could pretend it wasn't there. Look through the gaps in the fence, the wire hooked between your knuckles, and all that lay beyond was dancing white petals. Daisies, dozens of them. A brief fever dream amid the brick and concrete.

I'd last walked past it last a couple of weeks before, wandering back from a dinner that had been served up in a courtyard. It was a civilised thing to do on a Sunday night: meet with friends and crack open shellfish, mop it up with bread. Someone had taken a selfie, posted it online. This was a mark of our comfort, our accomplishments. These were the kinds of things my generation had been made to want: simple delicacies with like-minded people somewhere we could walk home from, even in London, on the first balmy night of late spring.

Josh and I headed home up the hill holding hands, and I pulled him back to look at these flowers. Sometimes it felt like a novelty, that this was what life was. A bit of an elaborate joke, of playing pretend. It felt both too good to be true and yet never quite enough;

always at a slight remove from what the roaring essence of life should be. Perhaps that was because it wasn't really meant to be this way.

Everything punctured after that, the air rushing out so quickly that it left me dizzy. Here I was now, taking in this rare patch of undeveloped scrubland littered with wildflowers and wondering where I would end up. How I had been in something that just didn't exist any more. If somebody mowed these flowers down, would they grow back the next year? Maybe we were just to have them for the few days that they drifted in the fading light before crumpling, weighed down with seed.

•

When I was a child, wildflowers were weaponry. We saw nature's offerings as something both prosaic and powerful, plentiful ammo to be deployed in the constant fantastical battles that defined our countryside upbringings.

Stickyweed was to be pulled down, balled up and tossed so lightly towards the victim that, ideally, they wouldn't know they had been targeted for several hours. They would be left to wander around unwittingly, the bright green barbs stuck to their T-shirt and covering their spine or shoulder or, best yet, their bum, for as long as it took for someone to point out what had befallen them.

Dandelions served other potentially punitive purposes. Come May, when their scraggly yellow flowers had blossomed into far prettier drifts of fine fluff, they became soothsayers. Those blowing the seeds off a dandelion head could divine many things with their breath, but mostly chose to establish whether two people

– often a nervous friend and either the most or least desirable boy in class – loved one another, or not. More potent horrors lay inside the weeds' stalks, though. Those encouraged to suck on the snapped stalk of a dandelion – usually by being promised a delicacy – will find instead hearty bitterness from the milky sap that had landed on their tongue, a grim taste that lingered and contorted the face, much to the glee of the perpetrator.

But the most cunning of the lot were the grasses. As the days lengthened, they would grow long and swaying, erupt into seed heads that held tiny spears and scatter bombs. We never knew their names, but we knew how to pick a good one – something with plenty of seeds but not too sparsely distributed. The opulent, fluffy ones were bold and advanced choices; rookies would be taken in by the shinier, spinier types, but these were too compact and would not deploy themselves well. Rather, something in between was needed – and it took experience to spot. In spite of spending the first chunk of her childhood in the suburbs, my sister learned this swiftly and was bolstered by knowing my gullibility and my desires well. She would choose her weapon, tell me to let her lay it across my tongue, grit my teeth and close my eyes if I wanted to know what it felt like to fly. Then, blade correctly placed, flying sensations suitably hyped up, she would tug the stem that emerged from the side of my lips and cackle as I felt the hard, dry seeds explode behind my teeth. I'd open my eyes and see her laughing as I spat out the seemingly endless supply of seed, removing it from inside my chops. A whole new kind of language left on my tongue.

I know these tricks because I suffered them plenty and rarely succeeded in dishing them back out – I tried to make Hannah

For all that, though, life was largely lived indoors. The village may have retained its most baffling and charming traditions – hog roasts and playing with sheep bladders and fierce rivalries at produce shows – but I was still a child of the Nineties, as enticed by technology and the siren call of the future as everyone else. I have an exquisite memory of a Windows 95 computer being installed in the study and a similarly crisp one of being shown how to access the internet a few years later. The possibilities of a life online surged through our generation and those around us like a tidal wave, and yet few could predict just how one would unfold.

As a teenager, I grew claustrophobic in the countryside. All that space, but no means to escape it. I lusted for the city, for London, for pavements and street style and a sense of danger and revelry beyond the worry of a too-fast car on an unlit back road. I felt stifled by the village's silence, the expanse of its skies and the sometime smallness of its mind. Meanwhile, our parents and teachers asked us what we wanted to be, chivvied us into becoming things, finding callings and careers and job titles. We'd parrot them, sparking the need and desperation to cast ourselves a future. I alighted on being a journalist, someone who made work out of play. I wanted my words on a page. And so I left for a string of over-growing cities. And I didn't think about the plants or the seasons or the cycles I had left behind until I came to realise how much I missed them.

•

When I first started to take an interest in plants, somewhere in my mid-twenties, it was to embark upon a journey that unspooled slowly. There was nothing showy about it. If anything, I kept it deeply hidden. To get addicted to gardening was considered strange and dowdy, a habit enjoyed by the elderly or the tedious. The steady, lingering gratification that remains after discovering a new shoot or unfurling leaf, of opening the airing-cupboard door to find a dozen germinating seeds pushing at the edges of a propagator, couldn't be captured in a photograph that would easily sit alongside the more common fodder of a millennial Facebook feed: three a.m. snapshots from a Hackney Wick club night or the views from a mini-break in Budapest.

And I didn't quite understand why I enjoyed it, either. I hadn't been raised to get stuck into gardening; I'd never before felt a need to study botany or a longing to visit public gardens. The trappings of gardening – slightly naff graphic design, an assumption of knowledge and a certain pernicketiness – still left me cold. All I knew is that it gave me a pure enjoyment I had not found elsewhere; not in London's bright lights, not in fashionable parties or hyped-up albums. To indulge in plants was to ask dozens of excitable questions about how and why the plants were doing what they did. I wanted to know how to answer. There was a silent, unspoken challenge about it all that didn't need to express itself anywhere beyond my own brain. And, unlike the other, more shouty propellants in my life so far (get the best grades, get a degree, find the perfect job, make a group of friends with whom to have the kind of fun that looks good on social media), there was no determination to gardening. The level of effort you put in did affect what came out the other side, but the causal relationship was a slippery one, one divined by elements beyond my control. For someone who had spent a very long time trying to push everything in the right direction, this felt like a constantly charming magic trick.

Like millions before me, I moved to London to find work. I adapted well. I found comfort in the noise and the anonymity, and fascination in the constant change. But a city is something humans made out of need, and the result has become difficult to live in. There is little space left for thought and reflection. Here, away from the hundreds of fine, tiny changes in air and earth and branch, strange demands were made of me. The city changes our priorities, forces us to compete in ways we never thought mattered to us: in terms of our income and where we go on holiday. More of us live in cities than ever before. The millennial generation – the one I belong to – flocked to these masses of grey and glass and steel, threw ourselves into housing poverty and clamoured for jobs in recession-scarred industries. We tried to shake off the expectations held by our parents while forging new ways of life; we wanted to do things rather than own them, even while attempting to buy a flat. We scrabbled up career ladders that led to futures that were kaleidoscopic, shape-shifting and impossible to predict. We tried to be many different things at once, got good at pretending even when we felt like we were failing at all of them.

We had been pushed away from the other living things we shared our space with. We grew plant-blind, ignorant to the power and the purpose of the greenery that we no longer knew how to identify. And we weren't the first: for generations, people have left the countryside of their childhoods for the fancy riches of the city. Eventually, the land claims us back. We find ourselves seeking it out, this restorative green space. We defy law and doctrine to grow things in soil that is not ours, making the dull beautiful to soothe both the hearts of the masses as well as our own. In the wake of the smut and the smog of the Industrial Revolution, Victorian authorities began to carve out space for parks, so that people could breathe from green lungs when their own became filled with soot. Later, when the frenetic pace of that century's invention left its children weary and worn-out, it was with garden design that the most cutting-edge creatives tried to find new freedoms.

Where do we sit among these generations? What in our indoor lives has come to shape our brains, our needs and wants? I found myself craving the brittle taste of them again, those unexpected grass seeds. I wanted the surprise of it across my tongue, something given, no matter how roughly. I sought an expanse, not necessarily of where I lived – for the city is large and full of as much wonder as it is frustration – but of how I thought. As I stared at those daisies, occupying the pavement for whole minutes as others quickly walked past, I realised I was hungry. Hungry for a kind of understanding, the kind of humble superpower that came with turning stickyweed into a gag, a fattened blackberry into an inky snack or a dock leaf into a remedy. It seemed that if I could only navigate the workings of these plants, to tune in to what made them bloom and shrink, that I could find a whole new way of living.

JUNE

THE FIRST HIT OF SUMMER in the city lands with the same high pressure that causes it. Brick walls soak up the unexpected sun, tarmac shimmers with the bake of it. We are sweaty, accidentally bundled up in our tights and coats and boots. A giant palm has been laid above us all and we celebrate by flocking outdoors, to the gardens and parks, to crack open tinnies in a million hissing gasps. We know it won't be hot for long.

People tend to forget how wet and showery June can be. A sunny weekend early in the month, oft declared a heatwave by certain newspapers, will usher the summer open – even though the solstice, the tipping point between light and dark, won't arrive for weeks. But rain will follow, it always does. It's the combination of both the surprising blister of heat and the runaway gurgle of persistent rain that allows the plants to grow.

Because June is fertile. There is a pause between the dainty abundance of spring and the heft of summer in its peak. In June, things are growing and gangly, on the cusp of riotous change. Hollyhocks spring up from the earth, looming on kerbsides.

Tree-lined roads appear to shrink as the boughs fatten with leaves. Grasses become wild and swaying, there to catch the back of knees. Roses explode in softness and scent, ready to become heavy with rainwater. There are so many buds that, after wind and rain, some end up on the pavement, offering a crunch under passing feet. Everywhere is green and teeming and eager with it, this burgeoning sense of new life. The solstice nears, tipping the world on its axis. It changes the shape of the days we fill with everyday things.

•

My life had been steady for a while. It was the start of the third summer in the same home, the longest stretch of time in one place throughout my twenties. That flat bore the weight of the seasons, looking, as it did, across the city from the fourth floor on top of a hill, capable of catching both dawn and dusk from the dining table. It would steam up in winter, condensation trickling down windows letting in the feeble dawn, water pooling on the sills. Storms would batter it. And with the bright heat of summer, we would open it up and let the day stream in until the silhouettes of evening painted themselves across the blushing walls. A brisk wind would rattle down the hallway and slam the doors at either end of it, interrupting the blanched calm of the place.

This was the ship that we commandeered, Josh and I. A gleaming white home that sometimes felt too grown-up for the stuff we had accumulated together, too polished for what bound us: adventure and appetite.

We'd fallen in love five years earlier, in the next neighbourhood along, over a summer of packed lunches in the park and walks along the Thames. Whole warm weeks passed before we kissed, a few minutes after midnight, next to the lions in Trafalgar Square. He was another thing in the new, lonely sprawl of London that I drank from, bitter and refreshing and moreish. After that, we were rarely apart, falling into a relationship without really knowing what one was. His early twenties unravelled into mine. Quiet and thoughtful to my rapid noise, he showed me care unlike any I had known before. I, meanwhile, tried to tug him out of every tight perimeter of his comfort zone.

We were each other's formative loves; the ones that blossom brightly on unsteady ground, root down between the cracks of youth and keep going in spite of unseasonal weather. And we had ticked the boxes of young adulthood: dancing until daybreak, travelling far, falling out. We stuck together in spite of illness and heartache, we learned to put another person first even when it hurt. We worked hard at it, this love. Patched it together out of fierce support and understanding, making do when we couldn't make it better. Our lives folded into one another, as lives in love often do. Human origami; we had become practised at it.

With time, it felt like we grew into something other and different to ourselves. We were connected by our ambition and determination to get the careers and lives we wanted, but also the things we had built together: impenetrable Escher castles of language and humour, spinning tales that could be reduced to code-like snippets. How we revelled in it, this secret, snow-globe world shut off to others. I'd never met somebody who worried more than I did until I met him. He made me seem free and easy in a way that others never had. But nor had I encountered someone so committed to looking after me, so strong in their moral compass, so uncompromising in their understanding of what was wrong and right, and so quick in the mind. I loved that he unspooled himself slowly, that to know him was like learning a well-earned secret. And so when we grew up before everybody else did, it didn't seem to matter so much because I was doing it with him.

The flat was a mark of graduation, of commitment entrenched in serious paperwork and legalese. Binding things. We were among the very lucky, very few home-owning millennials, and in London. The ones who bucked the headline horror stories, thanks to a mixture of inheritance, the generosity of others and a beyond-our-years maturity. Made of bricks and mortar and yet I treated it like an eggshell: a precious and often preposterous casing for our nascent lives. More a new toy that had been bestowed upon us than a place to live.

We tried to make a home that smothered even our young world-weariness in comfort, one that copied the Pinterest boards with Freecycle trophies. With time, the novelty of the place softened. We conducted normal life in it, sandwich-making, teeth-brushing. Took in a lodger to help us with the bills and slipped into different bedtimes. And I started to push beyond its boundaries and into the world outside, through the door to the balcony.

The balcony was my favourite bit of the flat. I relished the dinkiness of it – less than four metres long, just over one wide and flanked on either side by weatherbeaten Crittall-framed doors so small that people gingerly stepped through sideways while commenting, usually with a nervous laugh, that they might get stuck. Once I stepped through them, though, I felt a gush of freedom; to see the sky, to feel as though I was in it, was to breathe properly. My lungs felt bigger; there was more room to exhale.

Tentatively, I started to colonise it. I found myself spending more time out on that little sky platform. I wanted to bring life to somewhere that felt so gusty. I started with herbs – mint, thyme and sage – and crammed them, straggling and rootbound, into industrial-sized tomato tins rescued from outside a pizza restaurant. I drowned their poor fragrant bodies within weeks. I fell into a routine of leaving the house early on a Sunday morning and heading east, to Columbia Road Flower Market, with a £20 note. I'd bundle what looked nice into carrier bags and take them home on the train only to accidentally, goodnaturedly abuse them in all manner of ways. Bargain plants from Sainsbury's and Lidl offered horticultural training wheels. Things died, but others surprised me. It took me a while to learn that I must touch the soil before I watered to judge if the plants needed a drink or not. Instead, I'd just pour liquid love onto already drenched roots. I subjected tender growth to bruising winds. I saw height, plants growing tall if insubstantial, as a triumph rather than a sign of desperation for light or food, and when my plants bolted (going to flower, in order to make seed in a last gasp of energy before a premature death) I left them to bloom out of a mixture of intrigue and pride. And some were justifiably beautiful; even now, I will let rocket bolt quite happily: its delicate, windmill-shaped white flowers are among my very favourites. Just before they go over, I cut them from their stems and add them to a salad, savouring the novelty of their softly nutty flavour.

While I had grown up in the countryside, the granddaughter of two men who had greenhouses and vegetable patches, who would find solace in thinning-out and breach their upstanding morals by pocketing cuttings from National Trust gardens, I hadn't taken an interest in gardening until now.

It's not that I was averse to nature: my childhood was one of bike rides, field-conquering and den-making. But there were books to be read, pictures to be drawn, fleeting fascinations with friendship bracelets and dance routines. I was prescribed glasses at seven and promptly became obsessed with wearing them; the kind of child so reticent to play outside that my mother would threaten to move us all to a flat without a garden until I did.

When the seeds of interest started to sprout a couple of decades later, gardening wasn't really the done thing. It felt like the most pathetic kind of rebellion at first: no drugs or sexual boundaries conquered, merely the ground. It wasn't clubbing or brunch, a long weekend in Copenhagen or a group holiday to Koh Samui. People my age were expected to do many things, often all at once – travel, work creatively, party hard, present well and sleep with one another in ever more fluid ways – but growing things was never one of the socially prescribed activities.

And why would it be? The soil beneath our feet was an alien thing, something to launch ourselves off from, into the giddy stratospheres of post-millennium promise. We had been brought up by parents who witnessed the rise of supermarkets; those of us born in the last decades of the twentieth century were two generations distant from the people who grew to eat and enjoy. Front gardens weren't pruned in the Nineties; they were paved over. House plants were replaced with artificial flowers and potpourri. Conservatories, bike sheds and endless metres of decking took the space where greenhouses used to stand.

We learned the essentials of housekeeping – how to cook, clean and find vintage furniture on kerbsides – but those of tending to life outdoors became less relevant. Plants were superfluous. Even in the countryside they became merely the backdrop of a world constricting in its distance from other people. I craved asphalt and noise and the freedom granted by having a twenty-four-hour offy within walking distance, and so I found it. First in Newcastle; then, briefly, in New York; and finally in London, where it will keep me for a while, I imagine.

And yet, quietly, I was growing things. By June, jasmine was climbing gingerly up a drainpipe and purple basil was putting out leaves in spite of a shady corner. A courgette plant, still in its seedling pot, was flowering – even if the feathery reaches of powdery mildew would grasp its malnourished leaves soon after (courgettes, like most vegetables, need as much room and food as possible, and I gave them neither). The sweet peas I'd raised from pound-shop seed had been given their training canes. They would never flower, but in hindsight that wasn't a poor feat from such a stubborn-to-germinate plant. Recently I had been feeling inexplicably absent from the life I was living, as if I were going through the motions purely because that was what was expected of me. Fun, work, love: it was all muted, somehow. And yet here lay real thrill, in every unfurling leaf, every nascent shoot pushing above the surface.

I gardened with an abandon fuelled by curiosity, small successes and crushing failures. I didn't have the money to invest in my experiments, so I scavenged. I'd pot up trays of annuals (plants that germinate, flower and then set seed all in the space of a year) in a mishmash of rescued containers: wooden pallets, oil tins picked off the pavement outside curry houses and leftover plastic pots nabbed from the nursery. The second summer, I made my sweet peas clamber up a hideous wigwam I'd constructed from dead wood found in the park and twine. By the third spring, I'd used that same twine to truss a length of chicken wire down the brick wall of the flat for that year's crop to soar up.

And I did envisage that they would soar, even though they often limped. I was yet to learn about the distinctions of fertiliser, the hunger of the container garden, or the merits of a good feed. I was only just grasping the basics – of light, of shelter, of space – through my errors and some confounding online research. I aspired to grow it all, feeling nature's gentle confines only by pushing up against them: chard will not flourish in a small container, but sow an entire packet of mustard seeds into a series of them, and optimistically reuse the compost, and yes, leaves will appear two seasons later.

My knowledge accumulated like dust, without me realising or measuring it; there was more of it the next day than there had been the one before. It moved and persisted, changed with the seasons, gained with success and stilled with defeat but did not wane. And my enthusiasm mounted with it. I became idly hungry for the balcony and that which grew on it. It would stay more grey than green for years, but within that ragtag collection of pots and tubs and food tins were living things that existed somewhere in the gap between biology and my control. I'd linger at the balcony door, rest my forehead on the glass and stay there until, if it was cold, the mist my breath painted would cloud my vision. Josh would ask what I was doing, and I would always reply the same way: 'Just looking.'

Here lay endless fascination, but the balcony always remained my space. Others, including Josh, would come out on it sometimes in socked feet (I kept a grotty pair of flip-flops at the door, which I still refuse to throw out) and not know where to stand or look or put themselves. I was growing myself a cocoon without ever considering why.

Inside, meanwhile, increasingly became Josh's domain. I could be restless; tidying became a daily ritual as I attempted to instil my own sense of order on a space two people shared. I would spend whole weekend mornings cleaning strange corners of it, desperate to keep it lovely.

Between us, there would always be flashes of profound joy, the kind borne of years of familiarity; an hour of wild hysteria induced by pure silliness. But the rooms we occupied could also be a silent battlefield with opponent strategies orchestrated in banality: shoes in the wrong place and three-day-old papers never quite reaching the bin. At those times, when it was testy and heavy, the flat felt like an eyrie perched up on that hill, taking in all of London through its windows. A kind of cage. I'd look out over the river, to the east, where we used to live and where my friends still did, and wonder what I was missing out on.

I struggled to understand it, this frustration. The strange loneliness of living so closely with other people, of watching their lives stream in through screens. I had done all I was meant to do. I'd worked with the dogged fervour instilled in our education, education, education-entrenched generation. We were drilled by an exam system that turned success into a routine and near-perfect grades and good degrees followed. What awaited was a job market that demanded months of unpaid work for a faint whiff of a promise of steadier employment. I beetled through it, writing and working and picking up hangover cures, all in the hope of a career composed of bylines in newspapers and magazines people no longer bought. Of fulfilling the near-farcical dream of living off my words.

The jobs came along, in time. I became an editorial assistant at a buzzy new start-up and was greeted with a new laptop and a BlackBerry, which formed strange, invisible tendrils between me and the office. By twenty-four, I'd landed the kind of job that sounded good at parties – writing about pop culture for a broadsheet – and which was, in many ways, the realisation of hopes I'd hatched a decade earlier. When it was good, it felt like I was flying: going to gigs and festivals, writing about them and being paid for my imposter-syndrome-soaked opinions. But these were the rewards for a constant scrabble of trying to prove myself good enough over and over again, and having to swallow the casual rejection when I wasn't. To my mates, I had made it. To everyone in the office, I was just the new junior who hadn't yet grown jaded.

All that running around – from girlhood and school to university and internships – had led to days spent behind a desk. Cups of tea and eating over keyboards and gentle ascent of a pay scale that allowed me to inhabit, but never really live, in a city that only the very wealthy could properly afford. The hunger that had propelled me there changed and sated; I had either landed the picture byline or the print feature or the front-page puff, or I felt like I never would. Work became a thing to pay for the holidays we increasingly took from it, a daily battle with overflowing inboxes. Somewhere between the canteen and the keyboard, my ambition diminished. I stopped caring quite so much.

London became less of a holy grail than somewhere I had to end up in order to work. I hurried between work and pub or event and home, taking in the scale of it all only when crossing the Thames and blinking at all the lights. I found myself spending hours glued to the sofa, watching trash on Netflix, letting it engulf an evening, the next episode starting before I could reach to stop it. The loneliness beat inside me like a heart, and it felt like the only thing I could never tell anybody I felt. While my friends partied, or dated, or watched Netflix as well, on other laptops on other sofas in other parts of the city, I sank into these fictional lives flashing before me, tapping out messages on Twitter and WhatsApp and Instagram, until the only hours when I wasn't near an illuminated screen were those when I was asleep.

And maybe this was inevitable. My generation – raised on Global Hypercolor T-Shirts and *Gladiators* – had been urged to thrive online rather than outside. Gameboys, Playstations, Nokia 2210s and an adolescence navigated through MSN, these were the plastic-cased trappings of the first British teenagers to dwell in cyberspace and the last to grow up without it. Inside, we taught ourselves to touch type and torrent music, ramping up our demand for instant satisfaction with each boost in bandwidth. The glitch-filled song of the modem, repeated every time the phone rang, drifted away as broadband reached the provinces. I found out I'd got into university thanks to an email from UCAS before I'd managed to pick up my A-level results from school.

Digital time is a peculiar, distorted thing. News moves within minutes on Twitter; people can appear to hop from one country to another on Instagram. At work, I would silently compete with other people in other media companies to be the first to break the same nugget of entertainment news online. Everything had to be first, and fast. The fact that we had been at university without smartphones or Wi-Fi became a kind of quaint joke, as if we had teleported from the Dark Ages into the gleaming light of our devices. Such speed made other things telescope: I'd managed to arrive in my mid-twenties in an apparent idyll (nice boyfriend, lovely home, Instagram-worthy holidays) but without quite knowing how beyond the fact I was damn lucky.

It was as if I were floating in it, this strange accidental slumber, albeit happily. My lot was a good one really. Josh and I had an easiness, ways of pursuing our different interests away from one another. I delved into plants to a background of his affectionate mockery, and kept a similar disdain for his equally beguiling affectations. The contentment smothered me and drowned out any warning signals, a muffler against the sirens. I was certain of our future, that our lives would entwine and co-exist. Josh was just there – he always would be. We committed to the things that far older, more steady couples do: newspaper subscriptions, long-distance flights, bits of furniture we scrimped and saved to cherish. We joked about what we'd be like when we were old. I felt so safe in it, in him. We were, to my mind, an undeniable surety.

Days of life and weeks of work could stagnate or rattle by. Instead, I came to rely on watching the skies to mark the passing time. The flat came with a view from Battersea to Canary Wharf, everything in between small against the gleam of the Shard. Even that, though, was dwarfed by the painted air above them, which transformed with each passing minute. The clouds would thicken and the colours would shift in a silent performance that played out regardless of who was watching. With it, I learned to see how the sun would make its progression across the horizon with each passing day in the grasped moments of daybreak and dusk. Sky-watching from the balcony, losing hours to unravelling small mysteries, allowed me to place an ever-smaller version of myself within an improbably great system, one that worked beyond my control.

On the morning it all fell apart, the skies were clear. The kind of deep and unrelenting blue that leaves the ground carved up with shadow. I was looking at it as I mindlessly spooned cereal into my mouth when Josh walked into the room and told me he wanted to go on a break; that we needed to go on a break. Minutes earlier, I had eased myself out of his sleeping arms. These events didn't make sense. I couldn't process them, didn't want to. Perhaps he tried to explain, but I can't remember what was said; the words reached my ears distorted, as if he were talking underwater. The cereal softened in the bowl, slowly collapsing under the lapping tides of milk. I felt submerged by it all. When I came up for air, one sentence remained: 'I feel like I am falling out of love with you.'

•

The next few hours unfolded like tissue paper. My insides wanted to slump to the ground, let the day deal with itself and the next until this awful purgatory had run its course, but they were propped up by a determination to function as if nothing had happened. My generation's advocacy of openness, of talking about our anxieties and our mental health, existed mostly online. The propulsive need to turn up, do the job and leave late with a smile was far better entrenched. There were practical work things to be done, and so I got the grisly business of it over with – stood in the shower where I bawled with rage and confusion, allowing myself a few minutes of collapse – before putting on a brave face that I would carry, on and off, for the next year.

I daren't admit what was happening. It would let a fissure open that I wouldn't know how to close. It wasn't a normal work day – a shoot was taking place at the flat, and so I had to combine a broadcast face with the helpful unflappability of hosting. Every time one of my colleagues asked about the boyfriend I shared the place with I would pretend he hadn't just walked out, blood pulsing in my ears. At lunch, we took a break at the pub over the road and, while the others were fussing with mayonnaise and forks, I felt tears well up in my eyes. I gulped them down, hoped nobody had noticed. Inside, I was panicking. It felt as if my life had fallen off a precipice, landed in a mess on the ground and I was just looking at it, knowing there was no help to be called. The notion of existing without him walking in the door and saying hello was terrifying.

It was still light outside when he came back to pack a small suitcase and leave. I learned that I was not as loved as I thought I was, hadn't been for a small while, really. We muddled towards the decision not to contact one another, to give one another space to work out if he wanted to – if he was able to – come back.

When I wasn't fearful, I wrote it off as a blip, a small crisis, a necessary snag in the rich tapestry that would become our lives together. Perhaps, in a few years' time, we would think back on it, make quips about it at dinner parties while slightly rolling our eyes. That seemed like something that we would do. That was more logical, surely, than everything simply imploding. It was just a pause, and afterwards we would pick one another back up and become a stronger and happier couple.

But it had been exactly this kind of denial that had led to my surprise at Josh's departure. In the determination to be all things at once – a prominent journalist, a well-rounded twenty-something, good company on a night out, an adored best friend, a next-level girlfriend – I had started to deny that some of those things weren't working, that it was impossible to be all of them simultaneously. What we had had looked good on paper, had looked like what we had been raised to want. So when it finally arrived and turned out not to be right, I'd quietly decided to accept that this had to be right. While Josh had been struggling with the confines of our relationship and the ever-sharpening shape of the future I thought we'd have, I had been quietly shutting out reality.

•

By the next morning, the sky was muddled. Heavy clouds blew in, threw rain at the windows. I woke up in the bed alone, feeling only absence in its emptiness. My phone stayed quiet. I longed to find something on it, a message from him saying it had all been a horrible mistake. We'd always communicated fervently through our phones; waking up to text messages was something that started during my teens. With time, that transformed into waking up next to another body, and then only warm sheets, when that person had gone into the other room, because the Wi-Fi was better there. The silence was heavy.

June had grown cold and dank and with it the hours became slow, sullen things. I found the uncertainty – of whether he would take me back, of whether I'd be cast adrift – unbearable, despite how frequently I tried to solve it, working out countless different scenarios in my head, finding the only satisfying one a fantastical undoing of what had already happened. I hatched plans to leave the country, unable to consider a life in London without him. I just wanted to know what would happen, even as I stared down the crevice of an undefined break and imagined how the foreseeable would play out: spare rooms and sofas, damp flatshares, meaningless nights ending in tears, regret and more of the loneliness that was crashing through my brain.

Because nobody knew me like Josh. I was blessed with

had to sell the flat, I had no balcony. Without a balcony, I had no plants. I couldn't contemplate that nurturing or appreciating them could exist beyond these boundaries I had put up. It felt daft to be worrying about whether the parsley was going to seed. While gardening had grown from a dalliance into a compelling part of life, few things seemed compelling in that punching fist of fresh heartbreak. Without him, I lost interest in everything else.

But that combination of sun and wet was an intoxicating one for the plants. The modest – prim, even – collection of annuals and surviving perennials that I'd spent recent weeks putting in and tending to were reaching their zenith. A proud, violet spire of lupin stood against the wall, surrounded by bacopa, its froth of petals releasing a citrus sigh. In the corner, purple *Oxalis triangularis* had put out delicate flowers that weighed down featherlight stems. There were petunias and a rush of African daisies in garish pinks and purples. I drifted over to the glass door, rested my head against it, and stared out at the plants, not focusing on any of them. Suddenly, the flat felt very large and very quiet. Nobody there, now, to ask what I was doing. The strain of keeping on, holding the tears back, of sheer weariness, lodged itself on my brow, pushed heavily on my eyelids. A dull and persistent ache. It was difficult to make sense of the show going on beyond the door, of the bright colours and new growth strong against the falling rain. I made pacts with myself not to cry, broke them.

On the fourth afternoon, though, there was an awakening. I came home to find the grey, rain-slicked floor of the balcony interrupted by a dash of the new. Two fat, furry poppy heads had opened outwards, leaving petals as crisp, perfect and white as laundered linen. I inhaled sharply, they were such a surprise, defiantly gleaming against the surrounding gloom. Even buds I've been watching keenly for weeks – in the case of some, months – take me aback when they actually do bloom. It's almost like it happens in silence, when backs are turned or minds distracted.

Not that I had been watching. In the days that had passed, I hadn't kept an eye on what was growing, on which buds were swelling or which flowers going over. The magnitude of my upset, of the swirling confusion of Josh's departure, had seen me grasp for meaning in my rattling thoughts. I'd tried to impose order on impossible things, make sense of the unexpected. But here was something little – smaller than my palm – and unpredictable, and yet so right.

It made me realise that the plants didn't care. They didn't care if I was in love or out of it. They didn't care that I had stopped tending to them because I felt broken or that I had initially started because I sought to nurture something to feel settled, to fix something I had never even realised needed fixing. My state of being was entirely absent to them – of course it was, they are not sentient beings, not to contemporary human understanding, at least. And, regardless of what happened between Josh and me, of what people said or did to one another, the plants would continue to grow and bloom and go to seed and die back and grow again. Because that is what they exist to do.

The torrent of fear and upset and confusion paused, for a bit. Here lay the most reassuring notion I'd had in days. It took the wrenching apart of Josh and me and made it, just for a moment, seem small and banal. Heartbreak is such a ritual thing, an all-encompassing emotional takeover that happens to hundreds of people in every passing minute. Those poppies felt like a tiny miracle, a reminder that nature keeps going regardless.

I didn't make any great plans; there were no vows to garden my way to happiness. But there was a dawning recognition that unexpected changes didn't always have to be bad. That poppy was the beginning of a translation. The language of plants and of growing them was one I barely understood but found myself scrabbling to understand. I wanted to navigate the means of life that surround us silently every day. And when I began to try to translate it – clumsily, slowly – it helped me to make some sense of what had happened to my life; not just in terms of the break-up, but in what else I had expected from it and where I wanted it to go.

•

Flowers may be thought of as delicate and fey, ditzy things that inspire fervour in hobbyists and allow those with busier lives to pass by. But they have function and form and, as I was learning, a silent determination to flourish on their own terms. Women have long done the same, this steady upheaving of expectations and boundaries, and will continue to do so until our efforts are properly acknowledged. We have had to find our own ways into worlds and industries that we have been barricaded from, and that was as much the case with gardening as it was any other field.

Men imposed order on horticulture: they built towering greenhouses and amassed collections; devised the remits of garden design and created ways of putting plants in books that women were not allowed to write. Women were excluded for a good while, for similar reasons to those for which women were excluded from many pursuits: because their brains were deemed too small or delicate, because it was considered unseemly.

But we weren't always horticultural pariahs. In the nascent days of 'botanomania' – the eighteenth-century phenomenon inspired by the exotic plants that arrived on the drab shores of Britain from the edges of the Empire in ships otherwise stuffed with spices, tea and tigers – botany was considered the most appropriate natural science for women to study. The fresh air was thought to be good for us, the foreign plants gave us something different to paint and new herbal remedies to uncover: essential skills for a lady. By the 1830s, some knowledge of plants was up there with mid-level piano playing and polite conversation as skills a well-to-do woman should have.

What the patriarchy underestimated, however, was that we would want more than just surface-level plant admiration. That we would become avid and crucial collectors of plants, introducing species to the Royal Botanic Gardens at Kew and beyond. Women were responsible for bringing into the market some of the plants we grow domestically today: the delightful yellow fuzz of acacia and the delicate stripes of the cranesbill geranium were imported by a mysterious Mrs Norman of Bromley, thought to be the wife of a timber merchant.

I love the notion of these smart society ladies, dismissed by their fancy husbands, secretly forming a plant-importing network at parties. That they, too, were frustrated by achieving what society had told them was enough – a nice home, a well-chosen marriage and the right education and frocks to get on with a good life – and decided to push beyond that, to find something for themselves that challenged what they had been led to believe they wanted.

Others, though, were less enamoured. By the end of the 1700s, botanists such as Carl Linnaeus and Augustin Pyramus de Candolle started to classify plants and it led to a division in how some people saw women's enjoyment of them. John Lindley, chair of Botany at University College London between 1829 and 1860, was particularly evangelical in his mission to push plants beyond the drawing room and into the lecture theatre, where women weren't welcome (although he did later write the 'Bic for Her' of botanical publishing: *Ladies' Botany, or a Familiar Introduction to the Study of the Natural System of Botany*). And while the Botanical Society of London, formed in 1836, was the first scientific society to actively encourage the participation of women, the only woman to contribute scientific papers to the society – Margaretta 'Meta' Hopper – was published under her husband's name. Other scientific societies became a battleground for the sexes. Wars would ensue for decades before women would be granted entry to meetings – let alone fellowships – and allowed to investigate, discuss and explore plants like the men who founded these institutions.

But plants aren't grown in lecture theatres, and women (admittedly, usually the ones fortunate enough to have money and time) carried on growing and collecting regardless. The British botanical scientific establishment shut them out, so the women went elsewhere. Some would accompany their husbands (men who were governors of colonised lands such as India and South Africa) and go plant-hunting while the men were doing business. Others went on their own missions, commanded their own collections or defied convention by fashioning imposing gardens. And they studied plants, too, despite only being let into the Linnean Society fourteen years before they were given the vote.

For centuries, the collection, observance and nurturing of plants has been something women have turned to out of fascination, frustration and for relief. Bereavement, illness, scandal and heart-break have led us to make wonders of plants and gardens and forced us to find ways around the barriers that excluded us, even if it means sneaking into Kew Gardens at dawn, as budding entomologist Eleanor Ormerod did in the 1850s.

Women showed their potential as gardeners when they grasped at spaces that were often previously occupied by men. How wonderful that, when Charlotte Marryat – an American clever and confident enough to benefit from a loophole in the rules of the Royal Horticultural Society that failed to prevent women from applying and become the third woman to join it in 1830 – was left widowed, she spent her husband's inheritance on an entirely indulgent garden that held, among other things, a lake with two islands. Then there's Lady Dorothy Nevill, who, after being married off to an elderly cousin named Reginald Nevill in her early twenties, ignored social expectation and gardened with a thoroughly magical ambition. Sometimes it got her into trouble: when she attempted to farm silkworms at home, she ended up living with a plague of caterpillars.

A century later, women were still seizing land for themselves when vacuums in their lives appeared. Margery Fish undeniably loved her husband Walter, but he was both a fair-weather gardener (Fish would lay the foundations of her flower beds in the depths of winter) and a horticultural tyrant, deadheading plants only to leave the scattered prunings for Margery to pick up. When he died, she reclaimed what he had not allowed her, showing her true potential as a woman prepared to wield a crowbar to encourage the smallest of creepers to grow, rebelliously, in the cracks in her path. Within the gardening world, she is known as a writer and a pioneer, someone who moved the boundaries on what an English cottage garden could be. To Walter, one suspects, she was the secretary who became his wife.

When Sarah Lees was left widowed in her fifties – her well-to-do industrialist husband suddenly dying within a week of their twentieth wedding anniversary – she channelled her bereavement into improving Oldham, the town that had made her late husband rich. As well as funding scholarships and hospitals, though, she also believed in the power of green spaces. Lees founded a society dedicated to beautifying the squalid industrial town through parks, open spaces and flower beds. She experimented to work out which plants would survive the smog and hosted flower shows and cottage garden competitions to spread the benefits of growing things in a city that made human existence hard. With the turn of the twentieth century, Lees applied the same vigour to the suffrage movement, raising awareness of women's rights as she had the value of urban gardens.

Grief led the unmarried Victorian botanist Marianne North to explore the world. Her mother died first, leading her to accompany her father on his travels. When he died, bereaving her of the only travel companion she was willing to put up with, she continued alone. North, an innate feminist, had witnessed her sisters' marriages and decided matrimony was not for her, calling it 'a terrible experiment' that turned women into 'a sort of upper servant'. Instead, her life's work can now be found in Kew Gardens. A neat red-brick building belies the treasures within: eight hundred paintings packed in like photos in an album, created by North and housed to her specification. It is a relatively small gallery but kaleidoscopic nonetheless. Being there is like standing inside a jewellery box.

North began her solo travels at thirty-eight and did away with frippery. She didn't bother with elaborate frocks or the ambassadorial dinners her family connections afforded her, which could have made her years of travel more comfortable. Instead, she made her own way, telling people: 'I am a very wild bird and like liberty.' She roamed the planet according to geography and season – eighteen months schlepping across India, thirteen in Brazil – and painted hundreds of vistas that contained plants, such as mangroves, which had never been illustrated before and were on the imminent cusp of change. By preserving the plants in hundreds of paintings, rather than excavating them and bringing them back to the UK, North created a visual time capsule of landscapes that would vanish within years.

Being orphaned led Ellen Willmott to devote her vast wealth to lavish botanical expeditions, gardens in England, France and Italy, and dozens of gardeners. The list of things the unmarried Willmott grew is impressive and preposterous in equal part: 100,000 species, including one year in which she grew every variety of potato available to establish the tastiest. But what is more remarkable is that her knowledge was considered enough to earn her an RHS Victoria Medal of Honour in its first year. She was one of only two women to receive it alongside fifty-eight men.

The other was Gertrude Jekyll. Jekyll, who was born in 1843 and started collecting plants aged twenty, was so rampantly influential in terms of garden design and gardening history that her well-trodden story hardly bears repeating here. But for the uninitiated, she gardened because she loved to paint, and when her sight failed, she spread colour with flowers, leaving a trail of inspiration that still burns brightly today. What resonates most with me about her, though, is that Jekyll had no formal horticultural training. Nevertheless, she wrote about plants for *The Garden* from her late forties and a decade later released her first book, *Wood and Garden: Notes and Thoughts, Practical and Critical, of a Working Amateur*. That working amateur was also a working well-to-do woman who made her own money as a gardener. In doing so, she opened the floodgates for other untrained but eager and talented women to follow in her footsteps.

Jekyll and Willmott were good friends, and both gardened to the end. When Jekyll's sight had completely failed her, she could still recognise plants through smell and touch. Faced with a letter warning the repossession of Warley, Wilmott's beloved home and gardens, she went outside to weed. The horticultural heritage she's left behind, though, possibly says more about her. Miss Willmott's ghost, otherwise known as sea holly, is thought to have been named because Willmott had a habit of keeping the plant's seeds in her pocket to secretly scatter in the gardens of others. A quiet, spiky statement of Guerilla Girl gardening.

Just as women tried to smash the glass ceiling in horticulture, they also shattered horticultural glass ceilings to gain equality more broadly. On 8 February, 1913, suffragettes broke into Kew's much-loved orchid houses, broke some forty panes of glass and damaged the invaluable plants inside. Kew was as popular a tourist attraction in 1913 as it is now – luring some 3.8 million visitors between June and September that year – and the gardens' director had received warning of an imminent attack from the movement. The women acted in the early hours of the morning and got away with it, but poetically left behind a handkerchief and an envelope inscribed 'Votes for Women'.

News of the incident made global headlines and brought great awareness to the women's suffrage movement. Possibly spurred on by the impact of their vandalism, twelve days later two suffragettes were caught in the middle of an arson attack on Kew's tea house. Olive Wharry, twenty-six (although she said she was twenty-three at the time), and Lilian Lenton, twenty-two, had left cards at the site of the fire signed 'Two voteless women'. The court reports paint two outspoken, fearless females – Lenton went so far as to throw papers and a book at court officials during her sentencing. They threatened to go on hunger strike upon imprisonment, and they did. Lenton was released after force-feeding left her with pleurisy, but Wharry went without food for thirty-two days. Two years later, women gardeners were brought into Kew to replace the men who had gone to war. When they returned during the Second World War, the press insisted on calling them 'Kewties'.

Alice Walker dedicated a whole book to her mother's garden, *In Search of Our Mothers' Gardens*, for it was here that she found evidence of what her mother could have achieved had she not been black and living at the turn of the twentieth century before becoming a sharecropping mother of eight who took on seamstress work on the side. Walker's mother, 'so hindered and intruded upon in so many ways', still managed to plant 'ambitious gardens with over fifty different varieties of plants that bloom profusely from early March until late November'. She did this by gardening before leaving for the fields at daybreak and after returning, 'until night came and it was too dark to see'. The flowers she grew meant that Walker's 'memories of poverty are seen through a screen of blooms', that 'perfect strangers and imperfect strangers' stop to admire what she considered her mother's art.

Fury, justice and heartbreak brought women and plants together. Eventually, the paths they trod began to last, imprinting onto society. After centuries of being shut out, women helped to make gardening something everybody could participate in, and uncover the refuge and joy held in plants for themselves. This was the legacy I was digging into when I found myself dragged to the balcony window, just to look, just to find a small, unspoken salvation in the things growing outside. To colonise my own little space with play and creativity. An outdoor room of my own.

•

While I had been calmly ignoring the nuances of the shifting seasons for years – the different smells in the air and the hardness of ground underfoot – I had created my own annual routines for some time. The solstice was both my best friend's birthday and, often, the beginning of Glastonbury Festival, where the longest days would stretch out into streaky night for those who had entered into a temporary world where time no longer held power. Drum and bass could be found at eight in the morning, meditation at dusk. For five days in the Valley of Avalon, where mists rise and tors crown hills, there are tens of thousands of heartbeats seeking others kinds of rhythm.

I had been going for the past few years for work, another aspect of the job that induced envy in others but, in some kind of cruel Faustian pact, could only really arrive with a kind of cynicism seemingly requisite in the music industry. I'd grown from a fifteen-year-old obsessed with discovering bands and dredging meaning from their music into someone too busy, too exhausted, to hunt out new ways of listening. I struggled to care as much any more, and I was scared to admit it; it felt like a failing, a surefire way of being called out as a fraud. While I was there, I was to condense the airborne magic of that temporary city into copy while rushing between stages. It was invigorating: I liked doing things swiftly, I liked watching bands and I liked writing about them even more. But I could never escape the feeling that, even with the backstage passes, everyone was having more fun than me.

Nevertheless, I found it an easy place to love. A near-week in Glastonbury had become part of my annual calendar, days and nights beneath the sky, separated only by canvas and hood. I loved the rumble through the Somerset lanes in the coach to get there, that first glimpse of the impermanent city that lay beyond as the coach neared the site.

And this June was no different. A handful of days after Josh had walked away I was packing my rucksack, glad to be out of the flat for a bit, hopeful that the constant sensory overload of the festival would distract me, even for a minute, from my state of sad wondering.

But it was a difficult year. There had been heavy storms in the run-up; they slowed to a chilling, nagging trickle that permeated the days. My usual habits – joining Druids in their spiritual opening ceremony at dusk on the first night; ambling for a few solitary hours, taking in the mania of it all; racing to the Stone Circle to watch the sun rise on the Monday morning – were all hampered by the mud, feet-deep sucking stuff that must have held captive countless mobile phones, wallets, shoes and other detritus. Usually the festival site recovers (producing grass good enough for Worthy Farm's herd of 380 cows to graze on) in six weeks. But with the ground as ruined as this was, it took seven months. The earth needs time, space and a nurturing combination of sun, wind and fresh water to regenerate.

But it was also more difficult for me to shift my malaise and melancholy than I hoped. The previous days had been ones of shock and denial; the few friends I had told about it received the news in a wave of survivalist bluster; I shook it off in a gasp of self-deprecation, unable to dampen a chat at the pub with the extent of my upset. Here, though, in the rain-soaked outdoors, there was far less room to hide my feelings, especially from myself. The usual dysfunction and exhaustion that takes place over a muddy Glastonbury – twenty-one-hour standing-up days on three of fitful sleep – was compounded by the chaos in my head.

I missed him terribly. A full-bodied longing for the sheer physicality of him: the way the back of his neck smelled, the heavy comfort I found in the gap between his arms, the fact he always let me fall asleep first. I obeyed the rule that we'd drawn up, not to contact one another, but I wanted to reach out for him and did so in ways that would be delayed – sending endless thoughts into the air, postcards from the novelty festival post box. When I realised they had arrived, and I didn't hear from him, I felt a complete fool.

I was convinced I could feel the pain of it in my bones, in my heavy flesh and outdoors-blasted skin. My senses felt softened by it all. I could feel little beyond a curious senselessness that occasionally drifted into distant hope. Pride had meant I'd been unable to share the state of my situation with those colleagues I was actively working with; I didn't want to be treated gently, for I would just crumble. And I wanted at least one part of my life to seem normal, even if normality was, at that time, taking place in tents filled with people all seeking release and hedonism. And neither release nor hedonism was what I wanted, merely the security of knowing I could go back to a life I never saw a problem with. The notion of moving on was impossible, the idea of losing myself daft – I was already so lost. Instead, I concentrated on functioning, on being good at the strange task of working somewhere where everybody else was on holiday, because I felt so rubbish at everything else: at being a girlfriend, at being a lover, at being remotely wanted.

I used strangers like priests in confession booths, poured my heart out to people I'd briefly met in previous years in the hurly-burly of the press tent. Kind, middle-aged snappers who didn't mince their words and were there to take photos of James Corden and Alexa Chung in the backstage area, rather than, like me, hunt out secret sets and decide upon the most enlightening part of a performance. They shared their experience of love with me, offered stories that suggested it would all be all right in the end and, if it wasn't, well, then it was 'his loss, sweetheart'. Nobody could muster a different narrative: that I would go back with open arms and find Josh unwanting of them. That I would have to start again, and do so alone.

Meanwhile, the country was churning. While thousands of other people writhed to a DJ set under tree canopies, I stood with a stranger and, in the small hours, read on a glowing screen that Gateshead had voted to leave the EU. It felt like my generation's future had been stolen: a fearful, unimaginable blow, but one that in many ways felt like an extension of the mire I'd been living in. Little did we know of the years of turmoil that were being unleashed, of the uncertainty and havoc that would ripple through the country in the vote's wake and in how many ways all of our lives would be changed.

By the second half of the festival I had found a kind of release. After days of pretending to be all right, of trying to banish my tears and insist on upholding a kind of weird dignity to proceedings, I let myself cry. And cry with full abandon. Soft, gentle tears during impassioned anti-Brexit speeches made by P.J. Harvey and Matty Healy; full-bellied, ugly, snotty bawling as Adele played 'Someone Like You' and 125,000 people sang along. As Adele played anything, really, flares and flags swaying together in one great feeling beast. I sobbed at the sound of eighty people spontaneously bursting into 'Heroes' by David Bowie – a song that had always held great significance for the pair of us. I cried at the sheer unbridled beauty of Florence Welch running around topless on the Pyramid Stage.

The freedom of it, of weeping unabashedly in public in a place where nobody cared, and nobody knew me, was so utterly cathartic. I felt like a plug had been pulled out, that I was draining away feelings that hadn't just been pent up for the past week but for far longer. My denial was, gradually, being chipped away into the beginnings of a grief for what we had once been, and what I was beginning to learn we wouldn't be again.

As for what would fill that space, though, that remained unclear. Glastonbury offered all sorts of alternative walks of life for people, spaces to play and disintegrate, to shuffle into another mode of being. Here there was always another kind of party to go to. That year, I never felt like I was at the right one.

Instead, I found small solace in among the festival's permaculture garden. Peonies sat prettily in milk bottles, dodging the rain under the shelter of trees. A solitary bee made its way around a damp cushion of valerian flowers. While this ancient-feeling space had played host to hundreds of revellers during the solstice, once they had all gone home work would carry on: the compost heaps would rot and steam, things would flower, food would be harvested from the edible green roof of the educational hut in a corner. Glastonbury's permaculture garden was founded in 1989 and it has happily churned its way through the seasons ever since. There was a balm to be had by stumbling into it from the disused railway that strikes through the festival site, the calm industry taking place there. A living reminder of the other lives that take place away from the city I knew and inhabited.

With the short nights came clusters of promise and intrigue. Even though I was often fed up with it all, frequently returning to the tent for snatched naps after finding the trudge to yet another show too beleaguering, I was still captured by Glastonbury's night. There I found a new thing: The Sisterhood. We stumbled upon a caravan with a fluorescent sign on top. Inside, two brassy women sat behind a manicure bar purposefully giving the world's worst file-and-paint. My middle finger was roughly glossed and shoved in a pot of glitter and then I was pushed through a curtain. My friends, all men, weren't allowed in.

The caravan gave way to a dark room filled with chintzy furniture, tasselled lampshades and a low stage on which an angular girl in a white suit fronted a punk band. Women's bodies puddled into one another in corners, nuzzled into cushions and swagged curtains. I knew nobody, but found myself looped into a group of girls who took me in and then out to one of the Latin music clubs nearby. I danced with my new friends until around four a.m. Their names come to me like a faded patchwork, and I never did see them again. But for a few short hours I was able to bury my misery in this fleeting doorway of what could have been another life; the first instance in years of maybe tearing myself away from the cosy status quo that I had been slowly fitting into. Here was a dim and distant suggestion of what could be in an existence where Josh wasn't present, where I could be supported by women strong and confident and happy in themselves, and that I could be that too.

Enabled by the rare refreshment of a room where the rules had been drawn up by women, I started to think about where I stood inside the womanhood that had encroached on me during my twenties. I was resolutely hopeful that my relationship would recover but even that optimism couldn't ground me, couldn't release me from my denial and suspense. I was insistent I'd be taken back, that we'd have another opportunity to start again, even while I couldn't quite see how that would happen. The next night, I watched LCD Soundsystem's James Murphy mewl 'I Can Change' into a microphone, and it sounded like a prayer and a determination all at once. I felt that our fracturing relationship was my wrong to right, that my rejection had taken place because I had failed. I wanted to mould myself to make it work but something else was niggling: that I didn't have to be the thing to change, but other things could. After years of sharing myself with somebody else, I was on the cusp of being a single, thoroughly independent person for the first time since I was twenty-two. Perhaps I had to learn who she was as well.

Of all the signs, flags and slogans that pepper the Glastonbury site during its brief existence, one stuck out: 'What if we could live like this every day?' Part of the festival's magic is its fantasy, that it enables a semi-lawless existence built on freedom and agape, the kind of expansive love that is conjured by a crowd of strangers all belting out the words to the same song. Here, the structures of society warp and meld; the lust for power and money shrinks as people sleep beneath canvas and roam outside, slightly grubby, for several days. We could, of course, never live like it every day. But those words clung to me, rattled around my brain a bit. Made me wonder if I needed to push beyond the remits that I had been living within, take to the earth the same way those society women who broke with convention by hunting plants did. Regardless of what happened with my relationship, doing what we had been told, it seemed, hadn't brought either Josh or me happiness. As the whole country looked at a seismic break in its future, I was being forced to confront mine – it seemed naive to think that everything could just go back to how it was before.

On the coach back, I drifted in and out of sleep, knowing that I'd return to an empty flat and the heavy arrival of an uneasy reality laden with questions. The country roads through Somerset were drowsy with the dying days of June. Pin-cushions of hogweed nudged into wild grasses, white turned to pink; cranesbill geraniums buffered the breeze and hollyhocks towered in the middle of dual carriageways. I fell asleep as we left the country roads, white morning glory closing up shop, woke an hour or so later. Eyes opened to see fields of pale pink poppies swaying against the grey skies like spun sugar, a hit of surreality on the side of the motorway.

Once in the door, I let a week's worth of filthy washing drop to the floor and headed to the balcony to see a clutch of sweet peas in bloom.

version of myself. That maybe he had realised he had missed me, that his life was better with me in it, but it proved not to be. I didn't push for details of why or how Josh had reached this conclusion, but nor did I have the space to be curious. I just knew that I was no longer what he needed or wanted and he didn't think I would be again. After the words, he nudged towards leaving for some time, clearly riddled with guilt but also, I imagine, some relief of it being done. We walked away in opposite directions, and for the first time in five years I had no idea where he was headed.

I, meanwhile, got on a small series of buses, wielding a fluffed-up tissue at red cheeks, and headed to the countryside where my parents lived, carrying a bag I'd packed only in the hope that I'd be happily unpacking it shortly after, with him back in the flat, both of us embarking on a new stage of our lives together.

I didn't know how to explain it to my parents. There had been no grand fall-out, no deceit or betrayal. It was more as if my relationship had, at some point, been a lushly decorated, comforting room and, without me noticing, things had started to erode. Now I stood facing bare plaster, curls of damp wallpaper gathering around my feet, wondering how to piece it all back together again. The bits that were left suggested something wholly unrecognisable and, in that immediate aftermath, a lot worse. But what I didn't realise then was how, without Josh in my life, everything else would start to shift slightly, too. Gradually, space was opening up.

It felt cavernous at first. Even in the single bed at my parents' house, I'd wake up disorientated and wonder why he wasn't there before remembering he wouldn't ever be again. I'd try positing that he would change his mind in six months and we'd start again, but even then it felt naive. I messaged friends who had survived major break-ups and simply asked, 'How can I make it hurt less?' Hours of sheer nothingness – when I barely spoke, hardly registered the passing of time – would be followed by frenetic practicality, rallying together friends and emailing PRs for festival tickets, filling my near future with the constructs of fun. When I had stopped crying, I graduated to making phone calls, wandering around the house while talking to female friends who made me feel, for a few minutes at a time, like I was wearing my favourite T-shirt.

There was something strangely comforting in being held captive in the company of people my parents' age. I'd idle around the garden, cutting herbs and greens to put in dinner and picking up the fat, furry seed pods that had fallen from the wisteria. Everything moved more slowly. My godmother turned up, grasped me in a tight Lancastrian hug and whispered, 'Lots of love, Alice, lots of love,' fiercely into my ear. I was being coddled and adolescent and it felt like a warm bath. For days, modern, frenetic, showboating life just went on hold. Anna, a school friend whom I saw weekly in town, happened to be visiting her parents up the road, and she came over one night to keep me company in eating bags of weird crisps Mum had found on offer at the supermarket. It felt like we were sixteen again, and I loved it.

I continued on this juvenile streak when I finally got back to London. I started to undo the clearest trappings of our coupledom: unpeeled the photographs from our jaunts abroad that I'd Blu-Tacked onto hand-me-down shelving units, picked up his shoes from by the front door and shut them away in the wardrobe. I started to sleep on his side of the bed to compensate for his absence, but then just lay across it diagonally instead, taking up all the new room that had been afforded me.

I felt as if I had been grown-up for so long and now there wasn't really anything to be grown-up for; I could barely deal with being in the flat without him, let alone clean the place or keep it stocked with food. I took all of those years of feeling as if I was missing out, as if every other person my age was being more reckless with their hearts and their sleeping patterns, that they were having more fun, and poured them into weeks of hedonism.

I refused to stagnate, running instead through a chaotic performance of independent happiness found in bars, drinks, picnics and night buses. I laughed too loud; I drank too much; I loved my friends hard and told myself that was the same as falling asleep next to someone knowing they'd be there in the morning. All of it felt as sweet and brittle as Caramac. What giddy novelty came from drunkenly waiting for the bus tended to be extinguished once we arrived at the club, where I would look around at everybody dancing and wonder if any of them felt quite as broken as I did. But I'd carry on, pushing my body into vigorous shapes, hoping that the sweat would ease the pain. I spent far more time telling myself that this was right because it was what I should be doing, as dictated by some great and silent social lore, than because it was what I wanted to be doing. I took photos and posted them to Instagram in the knowledge that it would look as if I were having a good time, mostly because I suspected they would appear in his feed. Here was a self that I could offer up, even if it wasn't really what was happening at all.

But there was some genuine connection, too. I shook off that incarcerating pride that had kept me distant from my friends for so long. Released, for the first time in years, admissions that I had kept at bay, realising there was nothing shameful or silly in talking about the less-honeyed aspects of my life. Hard-boiling friendships came to quell the stubborn flow of feeling that my brain and gut allowed to run ceaselessly out of my fingers. I enjoyed lavishing myself upon people I had always kept slightly, politely, at a distance. I shared beds with them, watched sundowns with them, talked and talked and talked with them.

The sludge of June had turned into a balmy July, warm enough to spend weekends outdoors and catch a new lace of tan lines with each one. In hot, dry summers, the country gives up its secrets; lost villages emerge from empty, husk-like lakes. The foundations of a grand house, lost to fire and ruin, burnish through the scorched land like midday sunlight through a closed eyelid. At Chatsworth House, giant curlicued flower beds and paths laid down in 1699 presented themselves again. Like negatives, they appeared in the baked earth through their absence. And that was how I felt in those sunshiney, emotionally sodden weeks; as if bits of me that had been buried so deeply were coming to the surface with the sheer bronzed intensity of it all.

In return, these girls – and they were mostly girls – presented the shock and anger I was, and have largely remained, stubbornly incapable of; I've never seen sense in hating someone because they didn't feel the way you wanted them to. I enjoyed slipping back into their lives. Most of them were single and wildly accommodating, easy with laughter and generous in spirit. We picked over things and then, as the nights wore on, found more fun things to do. For minutes at first, and then hours and sometimes whole evenings, they took my hands and showed me other ways to be.

I told myself I needed it, this immersion in the unforeseen. Worrying has been my fear and downfall since childhood; when I was seven, I bought some Peruvian worry dolls from a hemp shop on a day out and they became a kind of strange deity in my bedroom for years. It was as if I were planning for the anxiety epidemic that would grip my generation as we grew up and realised the things we had been promised weren't to be; statistics show more than 10 per cent of us have been diagnosed with anxiety. I've always known that the excessive planning is futile – you can write as many to-do lists as you like but it won't stop your life falling apart on a sunny Monday morning – but I still do it constantly, on the back of envelopes and in the notes folder on my phone, for the air of calming satisfaction that it always brings.

So much had been left uncertain in the wake of the break-up. I was beginning to accept that Josh and I were over, but the more towering things – where we would live, what would happen to the flat, how we would divvy up the bits of life we had brought together – remained a mystery. Space is something that my generation is constantly struggling with. At seventeen, I could quote the applicant-to-space ratio of all of the university courses I applied to read; friends studying fashion at the better schools were told on their first day how few of them would make their final shows. There was a frenetic rush to get into student housing each November, as there just wasn't enough for all of us, and so it continued throughout the rest of our young adulthoods. Competition emerged everywhere: for those editorial assistant roles and unpaid internships, even. For the £500-per-month box room on the borders of Zone 3. It felt like everything had to be done in a hurry and better than everybody else simultaneously. There just wasn't enough space for us all, in anything, at anything.

So while I was well aware of how privileged I was to have the flat, it had become a situation. As we were both still paying the bills, were unable to rent it out and couldn't yet decide how to sell it, we decided it would be fairest to share, each taking the place for a month at a time, while the other person found another place to live. Josh had family in London, but I was to find a string of short sublets. I also knew how much of a challenge it would be. In London, rooms are leased and taken days before tenancy starts – even a ferocious, nervous planner like me couldn't put in any groundwork. I just had to wait for it to arrive.

I couldn't escape that tussle between change and control. I wanted to know how my life would pan out even as I pushed against the rigour of my usual routine. I'd taken an old university friend out to a gig I was reviewing on a Wednesday night; we didn't speak often enough for him to know about the break-up until I told him after the show, once we were headed into Shoreditch to stretch the evening out some more. He was always one for madcap schemes, rarely in the country for more than a few months at a time and always aghast at such mundane, constricting notions as, say, taking out rent on a room for a year. As students we would end up whiling away the afternoon in the deserted attic of a Lebanese restaurant, making our way through a plate of slow-roasted lamb, something that felt impossibly opulent in Newcastle, in November, when I should have been in the library. He was someone who would turn up at the latest possible second and engulf you in a spinning, elevated hug, shouting your name as strangers looked on, not someone to plan with or rely upon. Which is why, when I should have been going home and writing the review, I ended up kissing him in the bandstand that crowns Arnold Circus, a ring of Victorian red-brick housing blocks. It landed in a hurl of surprise, after hours that involved pizza from boxes and cider from tins and the most engaging conversation I'd managed in weeks.

I was blindsided by it, that daft, nostalgic kiss; foolishly saw it as an escape rather than a happy accident. He had been recently dumped, too. We were both a bit all over the place. But I couldn't apply any logic to it, allowed it to whip up a strange fantasy and frustration that, for a little while, usurped the melancholy that I had been waking up with.

I was so desperate for new realities to play with. Mine seemed difficult and annoying, mostly because it would only really get better with time. I had become so used to working for what I wanted, for the instant gratification that we are granted in so many ways, that I thought I could sidestep it completely. I entertained wild imaginings, both with him but also away from London entirely. An interview for a job cropped up in Copenhagen and I believed, quite firmly, that if I ran away from it all and moved into this

Pinterest-perfect Scandinavian landscape I would be fine. When I didn't get the job, I had to deal again with what actually lay around me: that I had to do this by myself, and that would take patience.

Somewhat mercifully, I wasn't alone. Relationships fell like dominoes that summer. Days after I had returned to London, started wearing clothes with fastenings and begun making an unsteady peace with my situation, Kate texted me to tell me that she, too, had been 'dumped'. Within hours, we were sitting in a park near the office and I saw a version of myself that had existed two weeks before, an inconsolable, grasping mess of dented womanhood. Within a couple more weeks a school friend had lost his lot, too: girlfriend, boat, dog, a dozen inches of hair. I felt the openness of their tender wounds as mine began to gingerly scab over, become raw to the touch. We sat on a rooftop in south London, worked our way through copious amounts of hummus, and I looked over at my neighbourhood knowing that it wasn't really mine any more, just the place where I fell asleep alone.

•

I cycled a mile or so up to Kennington, where Jamie was spending the summer on a street I'd always adored. I came across Courtenay Street during that first spring of living in London when I fell in love with the city by caressing it on two wheels. I had an A-to-Z that fitted perfectly in the front pocket of a tatty backpack and a mountain bike my dad had got from the tip. A minimum-wage internship at a magazine and London rent meant the bike and I covered the city's geography until the routes from Elephant and

Castle and Covent Garden, from Bank to the Old Kent Road, and Clerkenwell to the river were knitted in my mind. To learn them was to begin a tremulous relationship with the city, to stitch its streets together and lay them across my synapses.

Courtenay Street is one of a handful hiding behind Kennington's tower blocks. The imposing terraces line these arterial roads as if they've been lifted from another, more civilised city. The houses here are small, and gathered together by the water-icing curves of delicate white fences. On the neighbouring Courtenay Square and Cardigan Street, sweeping lead canopies drape over the doors between old-fashioned lamp-posts. If this was in Notting Hill, rather than smuggled behind the mucky thrum of Kennington Lane, it would attract a dozen Instagramming tourists a day. As happened throughout the city, it suffered during and after the Second World War, when the railings were taken for the greater scheme of things. By the end of the Sixties, though, a conservation effort had been put in place and now you have to have deep pockets to live there. Or, as Jamie had, convenient friends.

The house he was staying in had been entrusted to a friend by an old, recently passed artist who, I suspect, had been part of that bohemian swell that moved into lovely, run-down old houses five decades ago. The place reminded me of my grandfather's – a well-worn home of gathered clutter that had barely moved in decades. A postcard from Quentin Blake sat casually on a mantlepiece, battered enamel in primary colours stacked up on the kitchen shelves. The light from the French windows caught the dust and put a haze over it all.

But Jamie was also there because he, too, was nursing a heartbreak

dance every year and come back fighting the summer after. They don't have the scent of their annual cousins, the happy freshness of washing dried on a summer's day, but they make up for it in vigour. My favourite ones wrap themselves around the black iron railings of a Dickensian-looking terrace on Camberwell New Road. It is a house swimming in sadness, and yet every July a riot of cerise petals rises from the concrete, defiant against the fumes of a thousand passing cars. I've seen their sisters, the sea pea (*Lathyrus japonicus*) victoriously thrive on the pebbles of Dungeness – that bizarre, otherworldly triangle of coastline so dry it technically counts as the country's only desert. Perennial sweet peas will burst out of less raucous shrubs in front gardens and the hedging around the city's scrappier parks. They are one of the few flowers I will forage guilt-free (from public spaces, I hasten to add), because I know they have enough to spare, and it is usually a tussle to get to them – and worth every bramble scratch. Once home with stems in water, their blooms will stay preserved for nigh on a week, sometimes longer, the petals gradually fading to a dusty scarlet.

Lathyrus latifolius thrives upon the cruel confines of the city, but that doesn't stop urban gardeners from waging a yearly campaign with the more delicate *Lathyrus odoratus*. For me, as I suspect they are for many, the annual sweet pea is a country garden flower at heart. Its beauty lies in its simplicity; there are no showy spikes or dramatic trumpets here, just a few gently frilled petals pinched together like a handkerchief. Vita Sackville-West, great plantswoman and famed lover of Virginia Woolf, was somewhat more dismissive. 'Small, hooded, and not remarkable for any beauty of colour,' she described the true sweet pea in her weekly column for the *Observer* in October 1952 (the collected columns were later published as *In Your Garden Again*), referring to the wild Italian variety that was sent to these shores in the last year of the seventeenth century. It is from this 'humble little wildling' that the fancier and more fragrant varieties we grow in our gardens – Cupani, the Spencers, the Grandifloras – have arrived.

In the Fifties, Sackville-West was on the hunt for the Italian version, one that 'must be left to scramble up twiggy pea-sticks in a tangle and left entirely for picking, in an unwanted but sunny corner of the kitchen garden' (funnily, a description applicable to exactly what was happening in Jamie's borrowed plot). But at the same time, my maternal grandfather, three months away from becoming a father, may have been sowing his sweet pea seeds for the next summer.

I grow sweet peas because my mother grows them, and she grows them because my grandfather did, encouraging forth a mass of flowers three inches deep. Much as I tut at mine, mentally telling myself they're not as good as my mother's, so I've heard her tut at hers, muttering, 'They're still not as good as Dad's.' My grandmother would cut them and bring them inside and, I'm told, the flowers would fill their whole perfectly Sixties lounge with scent. In the kitchen I grew up in, cut sweet peas sat on the table in a blue-and-white ceramic jug, sometimes still with a confused greenfly attached, having not quite been shaken off during the short journey from the trellis beyond the kitchen window. Other times, the bunch would come from an old man named George who lived in a Victorian cottage near the end of the village and wore too-short trousers.

Because sweet peas are one of those magical annuals – like cosmos, which is best sown in May – that will reward routine savagery with more flowers. Granted, the stems will get shorter, especially if, like me, you are not growing them in a bed rich with well-fed soil. A container gardener friend – and flower fanatic – once told me that he grows sweet peas for nostalgia, in spite of the fact they hate even enormous tubs and will be spent by the start of July, while in the ground 'they carry on decently'. A well-cut, well-fed sweet pea plant will reward you with months of flowers if the season has been a good one; I've harvested blooms from my mum's on Bonfire Night. And this is because the only way sweet peas can go to seed – and thus ensure a chance of future progeny – is by flowering first. Those flowers wither and from their middles grows a pod, which if let be will eventually burst, scatter and possibly germinate.

This was what I explained to Jamie that Wednesday morning in July. This, and the way that gardeners interrupt that process. Days after those pale yellow buds swell, they push out petals. What ensues is a waiting game: how long the grower can bear it before cutting that flimsy first love of a plant down in its beautiful prime so that others will take its place. You can trick an annual like this for weeks, encouraging it into cycles of bloom and bust; tiny waltzes of control and patience, nurture and risk. How long to keep something alive, how much time it has to be beautiful and enjoyed before you stop it, unexpectedly, in its tracks, in the hope of encouraging something new.

Jamie's remaining flowers needed cutting, to preserve their petals before they shrank and fell, leaving another pod to join the masses growing plump above our heads. A rummage yielded no scissors, but we found a grapefruit knife and severed the stems with that. When I was done, I looked at him standing next to a yellow chair in a white shirt, holding a bunch of candy-pink flowers and breathing them in.

•

I grew up against music festivals. Watched my big brother – then just sixteen but seeming so grown up to me – standing in the crowd, pushing his floppy Nineties fringe out of his eyes as Radiohead beamed through the BBC. Glastonbury was still a wild thing then, years away from the unscalable metal fences and online photo registration that made ticket-buying a necessity only for the dedicated. People got mugged and beaten, the Travellers' Field existed as a dangerous other entity.

Some will say that Glastonbury has lost its edge now, but it's still the least corporate major festival in the country. The past twenty years have seen festivals become big business in the UK, a boon for tent-makers and welly providers, purveyors of glitter and flower headbands and street-food vans. I went to my first one – Reading, of course – when I was fifteen, and couldn't believe how many of my favourite bands were playing, all in a line, like soldiers.

But it stopped being about the music. People would go to dress up and gallivant, take a couple of days off work to escape the confines of their careers by wearing glitter in a field. The same artisanal mac 'n' cheese stalls and cocktail wagons would make appearances every weekend of the summer and others would come along for the ride: comedians and theatre troupes and literary salons. Banquets would appear in marquees, like a wedding where nobody knew one another. The whole thing was branded an 'experience', and therefore made it onto the list of things that millennials, who lacked objects and homes to be materialistic about, should be doing. To wear a spangly leotard and a fox mask and roam the woods in a haze of drink and ecstasy was a precious release made to seem necessary for those who were living the right kind of aspirational life.

I've been known to straddle ten of them some summers, and the summer I broke up with Josh was one of them. I'd reached a point in my career where the tickets and boutique camping were easy enough to come by, and because my friends and I had not committed much money or planning (it was nearly always a last-minute thing) we would pop in for just a couple of nights. It was fun, of course it was, and a total luxury. I still felt like I'd got away with something the minute I had yet another brightly coloured wristband tied around my arm and was released into the bunting-strung outdoors to do little but have fun. Dancing has always been a massive stress reliever for me, and I found release in staying up late and aimlessly wandering around on the hunt for moments of brief transcendence in the drop of a beat. There was something lovely in feeling the touch of cool air on your cheek in the dark of a summer's night, your body caught in a singing crowd. At times, these were the places where my brain would quieten.

I was lucky, and this was probably the best perk of the job I had grafted at for several years. But these tickets were a double-edged sword: you could go backstage, but never be seen to be having too much fun. To lose control, to get too excited, was embarrassing for everyone.

Such remits weren't put on those who had paid for tickets, who were there for a holiday. Or maybe they were, too. Festivals have become such a huge phenomenon that they come with social media profiles and YouTube accounts stuffed with videos of beautiful, sparkling people drinking champagne and dancing beneath the stars. There's no sunburn here, no listlessness, hang-overs or morning-after sadness. Even these constructs, offered as a means of getting away from the churning, grey reality of our lives in embracing a kind of freedom, have been packaged up into fantasies we're unlikely to muster without good lighting and some style direction. Was everyone else operating on this level, too, of not feeling quite gorgeous enough to participate in the hedonism that we craved?

I've always thought it was interesting that as the 'boutique' festival phenomenon grew the beauty of the location became integral to its appeal. Glastonbury, again, was a kind of progenitor of this – the Vale of Avalon is rolling and green, the mists come with the dawn to claim those making their way back to their tents. Now others transform and tout their woodlands into hidden glades for revelry to take place in, add in the stimulants of the club and remove the walls that sweat drips down. These ancient spaces – self-contained ecosystems of moss, mud, leaf and stem – become oddities for summer weekends. We sense that there is freedom to be found in these rarely touched wildernesses, but only tap into it while wearing fancy dress in the dark, flashing lights bouncing off the glitter we've pasted to our faces. Nature becomes a host for decadence, then we walk away and return to normality, barely noticing what we've left behind.

Nor are we the first generation to do it. The Victorians flocked to woodlands for years because it was fashionable. The trappings were different, but the principles not dissimilar: the notion of improving oneself was key, as was the idea of finding a higher grace. The most dedicated would seek out extreme locations and tread new ground only to take home the evidence to show off to their friends. The inspiration behind it all was one type of plant – the fern – and the Victorians nurtured a multi-faceted obsession with it for decades.

Ferns are the living dinosaurs of the plant world: fossils suggest they turned up 360 million years ago, while the ones that grow today popped up around 145 million years ago, and they've been calmly hanging out in places a lot of plants can't handle (rock crevices; shady woodland; wind-battered mountainsides; my balcony) ever since. As my mother remarked upon walking onto my little patch in the sky, 'I dig up the ferns that grow in our garden and Alice actively puts them in!' Ferns have been an easily ignored part of the landscape for centuries.

But the Victorians changed that. They found virtue in ferns and fell in love with their long-overlooked leafy minimalism. They saw the fact that ferns didn't flower as evidence that these were 'modest' beings. Botanists had only discovered how ferns reproduced – thanks in part to the brown, lumpy spores found on the underside of male plants' fronds – at the end of the eighteenth century, and despite their prehistoric origins ferns held an alien appeal because they took so long to give up their secrets. The Victorians took it upon themselves to unearth different species in far-flung corners of the UK and beyond, telling themselves that their pursuit was a noble one because it took such perseverance. Fern-appreciation was seen as instinctively godly naturalism for the light it cast on their creator's more unobtrusive efforts. It was no passing phase, either: in the 1830s a fern craze erupted in middle-class Victorian society that would linger until the beginning of the next century. By 1855, *The Water Babies* author Charles Kingsley invented a term to encapsulate the fascination with ferns that was sweeping households throughout the country: Pteridomania.

Pteridomania presented itself in ways that seem near-unfathomable today. Those consumed by the phenomenon – known as Pteridomaniacs – would pride themselves on the study of ferns and learn their ever-lengthening and often multitudinous names. There was a publishing boom in books about ferns ready to fuel those fans eager to learn. The main crux of the fashion centred around fern-hunting, in which fern fans would go on expeditions in search of unusual specimens that they would then bring home and keep in glass boxes, known as Wardian cases. Nona Bellairs was one of dozens of women who documented her Pteridomaniac travels. In 1865 she published *Hardy Ferns: How I Collected and Cultivated Them*, which details the three-month trip she spent in Scotland armed with two different trowels, an identification book and a 'large tin box with padlock and key' in which she carried her calico-sewn specimens for weeks on end.

Bellairs may have been forty-one when she published the book, but that didn't stop her from getting into perilous situations in the name of securing a species. She gaily rattles off an incident in which she prods at a 'beautiful [*Asplenium*] *marinum*' using 'a bamboo fifteen feet long, with a knife tied on the end' inside a sea cave. The tide was rapidly coming in and she had to be saved by 'a gentleman, a lady and a sailor' in a boat, who later 'gathered round the fern, feeling we could hardly admire it enough'. Bellairs was lucky, and evidently lived to tell the tale: others were less so. A Miss Jane Muers died in Perthshire in 1867 when the cliff beneath a fern she was collecting gave way under her feet.

There were whole excursions and trips organised with Pteridomania in mind. Thomas Cook & Son were among the tourist companies who recommended stops at fern gullies and ferneries on their itineraries. In areas of abundance, such as Snowdonia and Windermere, enterprising locals took a more grassroots approach, accosting the hoards of Pteridomaniacs (who had arrived by the recently built railway) with fronds they had dug up and presenting themselves as guides who would show the new fern-hunters where to secure their bounty.

Curiously, Pteridomania was perceived as a particularly female hobby. There were special designs of kilted dresses made to allow the Victorian lady greater ease of movement while fern-hunting, and younger women clung onto finding, identifying, keeping, classifying and drawing ferns as teenage girls would members of pop groups a century later. In coining the term in *Glaucus; or, the Wonders of the Shore*, Kingsley also showed that he was gently approving of the movement:

> You cannot deny that [your daughters] find an enjoyment in it, and are more active, more cheerful, more self-forgetful over it, than they would have been over gossip, crochet and Berlin-wool.

Charles Dickens, however, was less enthused. In 1862 he refused his daughter a Wardian case (which cost the modern equivalent of between £200 and £500, the same as a weekend at some festivals) after 'carefully cross-examining [her]' and coming to the conclusion that he did 'NOT believe her to be worthy of the fernery'. He wrote in a letter to a friend:

> I am quite confident that the constancy of the young person is not to be trusted, and that she had better attach her fernery to one of the Chateaux in Spain, or one of her English castles in the air.

Towards the later years of Pteridomania, the women in Edith Wharton's *The Age of Innocence* were more fortunate, and nurtured their fictional Wardian cases quite without opposition.

Perhaps it was just that women, so long prohibited from learning about plants – let alone being actively encouraged to discover them – were enjoying the new thrill of the botanical chase and the academic rigour that came with it. Women had been given permission to tend to houseplants for nearly a century; in the 1770s, Josiah Wedgwood did market research for his newly fashioned bulb planters among his female customers. With the dawning of the Victorian age, women were gradually being allowed to decide what plants, and in what arrangements, they kept in the house. Ladies' dainty fingers were considered better tools for tending to the proliferation of window boxes and indoor potted plants that sprang up inside the newly built modern homes, which accommodated them with larger windows. With a surge of urban building, these new, mostly female, city-dwellers treated houseplants and the ability to nurture them as a status symbol. Kentia palms, popular now because they are wildly tolerant of relative neglect and will put up with the low light of a dingy hallway, were equally beloved by Victorians, who found they would survive in their soot-filthy, gas-filled living rooms. Gardening books and magazines aimed at women proliferated. In 1842, a Mrs Jane Loudon wrote in *The Ladies' Magazine of Gardening* that:

> I have no garden, but as I have a large balcony, I have many greenhouse plants, which look very well during summer, but which give me a great deal of trouble in winter. I have been obliged to line all my windows with them, and I have flower-stands full of them in all the living rooms; but there are still many which I am quite at a loss to dispose of.

Given the craving for greenery indoors – and the frustrations that arrived as a result – it is hardly surprising that these women, already largely banished from gardening outside or even in their own greenhouses (considered a more manly pursuit), found such glee in being given permission to run amok in the woods on the hunt for ferns, let alone with one another. There's a sisterly joy to be found in a photograph of a jam-packed audience of women in high-necked blouses attending a 'Lesson on Ferns' at Pocono Pines, a Pennsylvanian mountain resort, in 1900. Several species were even named after the women (such as Miss Beever and Mrs Bolland) who found them. Granted, it wasn't all good: it's difficult to read about the rampant removal of ferns from their natural habitat only for them to sit, sweat and suffer in glass boxes in dark drawing rooms. Even a fern lover as proclaimed as Bellairs writes concerningly about the species she 'ruthlessly' digs up and mangles into her tin box, only to disown once she gets it home.

> I never have made you live, and I fear I never shall [she says of the *Botrychium*]. I am obliged to own that when I have found it I have never kept it alive for any time . . . Still, it is worth hunting for.

The term 'fern-robber', referring to those who pillaged with little care or understanding of what they took, lingered long after the craze died out at the end of the nineteenth century.

But for all the breathy excitement that accompanies Pteridomania, there are elements that feel vaguely familiar a century and a half on. Photographs of women perched in Australian tree ferns recall the artful shots of palm trees and bikinis that smother Instagram. The fern albums that documented Pteridomaniac travels and sat pointedly on coffee tables are the equivalent of our well-curated social media feeds. Ferns proliferated inside the middle-class Victorian home because for the increasing numbers of people who rented, they could be moved in – or out – with ease, just as the city-bound Generation Rent have clung onto houseplants as a vibrant, living thing that will stay with them beyond the precarious remits of an unkind landlord. Ferns rapidly became a social signifier. By 1840, botanist Edward Newman had remarked that fern cultivation 'is no longer confined to the botanist and horticulturalist; almost everyone possessing good taste has made, more or less successfully, an attempt to rear this tribe of plants'. And it spread into botanical prints and homeware, too. Just as palms crept onto foliage-clad wallpapers, prints, cushions, sofas and crockery in the mid-2010s, so Victorian fern ware became an interiors trend all of its own, and the skirts of fine dresses were embroidered with fronds.

When new trends emerge, the passion that thrust them there lessens. As the sad succulents in Urban Outfitters today belie the passion of the cactus fanatics who brought them from the desert to the domestic market in the 1930s, so Pteridomania at fever pitch seemed to forget about the early fascination that inspired people to get out in the clean country air and happily stumble across ferns in the first place. Edward Newman, one of Victorian England's first fern experts, only got nerdishly into learning their names after he was prescribed 'three months travelling about' for anxiety. Perhaps Henry David Thoreau summed it up better in his diary:

> If you would make acquaintance with the ferns, you must forget your botany. Not a single term or distinction is the least to the purpose.

It's advice that resounds far more strongly than the notion of fine ladies '[perambulating] country lanes with a gentleman or a footman to carry their basket and trowel'. I didn't have much botany to forget by the time I sank into that festival in mid July, but I wasn't applying what little I did have, either. And yet, I knew that the outdoors was what I wanted.

I had not been immune to the houseplant trend that had slowly begun to gain social cachet among people my age. The rustic sprawl of fiddle-leaf figs had started to pop up, propped next to sofas, in interior design magazines; *Monstera deliciosa*, known as Swiss cheese plants to our parents, gained a new modernity on Instagram and Pinterest. After decades in the wilderness, houseplants were making their way back indoors, into fashionable, stripped-back cafés and the bare minimalism of our homes. As the Victorians had before them, millennials were grasping at a nature that had been removed when they moved to cities and were pushed into small, expensive flats paid for by jobs looking at screens. We, too, made plants into a desirable thing; houseplant-laden homes filled the pages of aspirational websites dedicated to small urban living.

If Victorian teenagers wanted Wardian cases, we craved small glass boxes of our own. Terrariums – sealed glass bottles that contain their own tiny, self-sufficient green ecosystem – returned to fashion. On Instagram, we liked images of glasshouses and botanical gardens, lush jungles encased in a transparent cathedral of their own, in our thousands. These are worlds where control is key, where the conditions are kept static with the pure intention of nurturing life that couldn't exist outdoors. Ironic, yes, that an environment so embodied by captivity would be the ones that made us feel free, but the values that my generation has come to revere – of experience over ownership; of pushing beyond expectation out of sheer survival; and of 'authenticity' – could be found in these bastions of artificiality. When all in our lives is so transient, these glasshouses contain worlds where time is slowed down and marked only by nature's processes.

I kept houseplants in the flat, too. There was a rhipsalis I'd picked up in a charity shop and my windowsills boasted succulents that I'd bought in the nearby nurseries. The bathroom housed a fruitful aloe vera and the living room windowsill was variously scattered with low-maintenance haworthias and kalanchoes that I had received as gifts. At work a light-deprived cactus perched by my computer and in the mornings, when I got in early and the office was deserted, I'd enjoy watching the light cut through the foliage of the oxygen-starved palms that were deposited around the place.

But somehow I struggled to connect much with these plants. I'd accumulated them out of good fortune; often, they had been gifts – there was a short-lived fascination with putting succulents in goodie bags, PRs would send them through the post. It seemed churlish not to bring them in and watch them grow, but I was mostly ascribing to fashion. A succulent would stay much the same winter or summer; in our British homes they are small and tame, stifled versions of what they would be in their natural habitat.

The plants that moved me were the outside ones, those that persevered regardless of the weather. Those that existed to contribute to the greater world around them, that fed the bees and shed their leaves and threatened death in winter only to surge back with a long-forgotten vengeance with the promise of spring. I'd been pushing to be outside for months, to move beyond the boundaries I'd unwittingly raised around me; the notion of what my life would be. I'd lingered restlessly indoors out of not knowing how to escape it. Maybe that was why the air on the balcony gave me better reason to exhale; maybe that was why I found retreat in the walled gardens over the road.

There was something deeper behind my gallivanting about, beyond the notion of doing it because I could and had nobody to be around for. I longed to be outside. I'd been needing it for months, maybe years, and it was the same thing that lit me up on the balcony – only I wanted more of it. I wasn't just escaping London and all the memories that lay there, I was looking for something to root myself into.

This unspoken craving to be out, to be held captive only by the clouds, motivated me in the same way that music had when I was a teenager. I found solace and meaning in lyrics, would etch them in biro all over my school books and the bag that contained them. There was release in jumping in time to a drum-beat. I'd go to gigs on school nights and turn up when the doors opened to stand, stone-cold sober, at the front and wait for every support act to play. I caught an energy from the reverberation of a PA system, from watching someone throw themselves around a microphone, that I couldn't find elsewhere.

Over the years, though, that faded. Perhaps it is difficult to maintain such raw enthusiasm for something when it becomes a part of your job, to be scrutinised and calibrated and assessed for benefits other than how it made you feel. Or maybe I just grew up, grew too old to worry about being among the first to discover a band. No matter how many festivals I went to, how many bands I saw, the number I would see that made me feel that adolescent spirit and lust would dwindle. It was an apathy that was really difficult to own up to.

Still, though, I went through the motions. The headliners had finished, the revelry was underway. The crowds, increasingly lairy on plastic-cup booze and the early ignition of amphetamines, surged into the woodland. There were different options: tents of people dressed up as robots in tinfoil and cardboard boxes; tents of people wearing headphones, the sound of their singing and stamping alone filling the space; tents of late-night comedy and bawdy laughter; and then, our destination – a dance floor defined by the space it occupied between the trees, a clearing in the woodland before a stage where an anonymous DJ churned out beats. Rise and drop, rise and drop, the crowds pushed towards the speakers, the electronic thunder shuddering through their bodies. Mine joined in, my brain elsewhere, distracted with the loneliness that comes from wondering if you are alone in not feeling anything at all.

But then there was a different rhythm, of rain through foliage. It would have been loud without the music; it would have sounded glorious. Instead, it was silenced and had to be felt. Drops of moisture on my forehead, down the bridge of my nose, running off the backs of my knuckles. Hoods raised, faces tilted upwards, the dancing continued – the trees splintered the shower, forced the water to run in unlikely patterns. Eventually, the rain reached the forest floor. Beneath my sodden trainers damp, trampled earth stirred, the bracken nudged into my limbs. The smell of it – long-worn and riddled with life – besieged my senses, rattling up my nose and down my throat, rushing to my head and hitting me with a primal familiarity I hadn't encountered in years. It was the childhood rush of running through the woods, the unbothered calm of a dew-laden morning, the unmistakable energy of freedom. I was shaken by a physicality I hadn't yet known I'd been searching for, the same shunt that music used to give me. It was an escape that I'd been trying to create with consumption, not realising that I needed to let go of, rather than take, things to achieve it.

I stood still as the beats juddered the earth beneath my wet feet, watching my friends dance and hula-hoops cut the air into chinks of light, absorbing a space that had been going through these quiet, complex processes for millennia. The bracken (*Pteridium aquilinum*, a fern so common and invasive few collectors bothered with it) reached several feet high and formed a heavy pile over the banks of the wood, giving off a scent so fresh it offered a whole new world of breathing.

The woods kept drawing me back for the rest of the weekend, and I'd drag my friends with me. During the day they were less riotous and more of a sanctuary of dappled light and gently bobbing fronds. I sat on logs and bark chippings, largely ignoring the constant parade of acoustic guitars on stage and chatting with friends. I was jerked back to London after stumbling on an Instagram upload, which showed that Josh had spent the previous night in a stylish flat inhabited by some beautiful fashion stylist I didn't know, but I was with Kate, who knew to pour scorn on it and suggest we go and swim in the lake, instead. She was a balm for that, a potent combination of no-nonsense Yorkshire lass and assertive Westminster journalist who would conjure the truth from a person with ease. After throwing ourselves, squealing, into the water, we picked over our broken hearts as the lake cooled our skin and heightened our awareness. Comfort came when we bared our souls, our respective heartbreak softening the isolation we rattled around in. We wound up back in the woods, wet hair resting on the back of our necks, water slowly evaporating from sun-warmed skin. I could smell the pondweed on it for the rest of the day, heady and green. Here lay a space of escape that didn't demand anything, that I could come and relish away from the confines of my tarnished home, weary mind and idle wanting. Like those women before me who escaped their smart city apartments, laden with plants, to find purpose in the wilds beyond, there was something in this soil that gave me satisfaction.

In the end, I left the festival early. I'd had enough – for a few weeks at least. I no longer wanted that artifice, the construct of a good time. They weren't working for me; there was little satisfaction to be found in these gatherings of strangers. I enjoyed the feeling of leaving the dregs of the party in a double-decker bus that rambled through country lanes as an airy sunset filled the sky. I suspected that what I had encountered in the woods would exist in London, too, if I kept an eye out for it; that the same determination of the plants that had thrived there would lie in those that called the city home. It was a determination that rang true and deep, that existed even in the depths of my despair. It was what would ground me.

AUGUST

Roots are the subterranean skeleton and stomach of a plant. They offer it stability, and they bring it water and minerals. When the plant is working they are its pantry: they keep the energy the plant has generated as its very substance. And they are the first sign of life; a germinating seed will put out a sprout – one of the finest banal sights in everyday life – but only after the root has taken hold first, reaching into the depths to lay down foundations. Roots prefer the challenge of working into well-packed soil; an uprooted plant will settle quicker if planted firmly, with no large air pockets separating its root fibres and the earth in which it is to grow.

There are different kinds of root. Taproots, which burrow steadfastly down into the earth; the neatly uniform fibrous roots; and hulking, hungry tuberous roots. Creeping roots are the type that you're most likely to trip over, hiding under the shed leaves of a centuries-old oak, busting through the ground having grown old and tough. And adventitious roots are the optimistic ones, which spurt from green stems far above the soil just in case it becomes unearthed.

It's funny how humans have adopted the notion of their roots to mean their heritage. The family tree of any given plant is something addled by botany and curious human hands and minds; we will cross-pollinate and create new varieties. We call plants by confusing, different names (an asparagus fern isn't a fern at all, but part of the lily family – and it has tuberous roots) and remember them by the people who discovered them. Clues to a plant's heritage lie in many other parts of its being than that which sits beneath the soil.

But, like plants, we attach roots to places. Mine are not exotic, but I can recall them in gardens. A pond and patio in Berkshire, followed by the rectangular front-and-back lawns of my first proper childhood home. I was captivated by the imposing, fluff-headed sway of the pampas grass in the drive, which my parents swiftly ripped out. I'd lie under the tiny cerise flowers of the May tree (a pink hawthorn), named, my father said, after the month it bloomed.

My adolescence expanded in a long, skinny acre of long grass, punctuated by a light-sapping yew too old and superstition-laden to fell and apparently objectionable oak trees that I always liked. In the mornings, it would be covered by a mist that would creep in over breakfast time, leaving dew and freshness in its wake, forming the backdrop to our kitchen table. My dad's chair sat at the far end of the table. Growing up, the most exciting thing I saw from the window was a pheasant, which I found staring me in the face from two feet away upon opening the curtains shortly after one midwinter dawn. But Dad would spend hours looking out, although he frequently only allowed himself minutes, absent-mindedly tapping his wedding ring on his mug and looking at the green beyond. Maybe he was building mental to-do lists – he'd often mutter about the neighbour's eucalyptus – or going through the day ahead, granted space by a backdrop that changed in tiny ways with each passing day. No matter how sullen or sleepy I was in those hastily grabbed breakfast minutes before the school bus, Dad would offer me the best seat in the house – the one that looked down the garden. He would give up his, say he was done and potter around, stand up and look out of the window instead. And I, flicking through the paper, staring at the telly, never really noticed that small kindness. Instead, it was something I learned to do without realising. I'd come back from university and warm my thighs on the radiator beneath the window (that house was always cold), rest my head on the glass and watch my breath steam up the view of the garden.

I came to realise that I'd laid my life out this way in the flat Josh and I shared. I'd put the table parallel to the balcony, against the window, to sit at the far end of it for eating and writing and reading the paper. The view to the balcony, through that two-foot gap in the wall, became the backdrop against which breakfast and dinner were consumed and work and relaxation took place. Without design, my favourite seat in the flat had become the one that allowed me to look directly into that box of pocketed life. The only times I didn't sit there were when I had guests over and I, like my father before me, would secretly offer them the best seat in the house. I'm not sure they noticed; often they were too taken with the view of the Shard or the wine or the gossip. But I could reason with that – it takes time to recognise it. Although, thinking about it, I broke with tradition one other time too: on the night after the break-up, I spent those sodden, sleepless hours in the chair opposite, with my back to the balcony. It was as if watching the sun go down on my refuge was almost too much to face.

While I spent less time in them, my grandparents' gardens perhaps shine more vividly in my memory than the others. Grandpa, on my father's side, had a sprawling square of a plot behind his Victorian home, with hidden corners, a mulberry bush and a once-immaculate vegetable patch that he kept filled with potatoes nearly until the end. It was his utter joy.

Number 12 Albert Road was built with a greenhouse attached, and even into his late nineties Grandpa would manage to navigate around the piles of empty plant trays and bags of compost, much to the increasing horror of his progeny. Growth from outside and inside would commingle, pushing up against the glass and through the cracks, nature blurring the boundaries of infrastructure raised to control it. I was about seven when I was allowed into the greenhouse for the first time. He had a few *Schlumbergera* on the go, and sent me home with one; I managed to coax it into blooming, gaudy pink flowers from those chunky segmented stems. A decade later, I'd go and help him pot stuff up in the greenhouse. It was quiet, calmly slapdash work, full of the sense of a job done well enough.

Grandpa gently introduced the habit of an impromptu guided walk around the garden, something that took me time to understand, watching from the house as he and my parents admired the beds. Now, my sister and I do the same in hers – it's never suggested, it just happens. These informal inspections have a meditative quality; they allow the gardener to keep an eye on things – and learning to look properly is one of the most vital skills in gardening. Grandpa spent his last evening showing his closest friend around the garden, which had just come into the sort of lush growth that accompanies a warm mid-May. We were told that, upon admiring it, he contentedly concluded that he 'could go now' and, a few hours later, aged ninety-seven, he did.

If Grandpa had an eye for the botanical, then Grandad, the proud Yorkshireman who fathered my southern mother, was a grower. When I smell *Pelargonium* leaves – fuzzy and wildly underrated – and new tomato foliage, I am transformed to being small, sunshine-coated and being shown how things grow. There was a lot of unspoken pride in that greenhouse, as well as the stuffy green smell of busily managed nature in a little space. Outside, there were carrots that we would pull from the ground and then be ushered into the kitchen to wash the soil off before crunching them down, tasting the earth in each fevered bite. Years later I found a photograph in a box of snaps that didn't make it into the albums. It must have been early summer: the bean leaves are curling up straight bamboo canes on the right, next to carrot tops, leafy potatoes and, on the left, a mess of sweet peas is scrambling over a chicken-wire trellis. A hosepipe snakes over a seedling tray and on it stands me, aged two, round and grinning and wielding a trowel the size of my leg.

These are, in part, my roots: memories held in time and place that cling to the plants I grow now. Perhaps the origins of my botanical interest lie yet further back: a woman named Louisa Elizabeth Allen, my great-great-grandmother, painted flowers. As did Gladys Millen, the sister of my namesake – my great-grandmother, whose blue eyes and sharp tongue I have also inherited. We are a jumble of so many things. But if we reason that people can have roots like plants, then let's reason that they can be uprooted, too. And in August, I was on the move.

It was the first of several over the next half-year, out of what was once home and into more temporary residences. Over the past six churning weeks I had seen that there was some kind of life after it all; I'd developed an affection for the once-bitter taste of my independence. It fizzed like sherbet in my mouth, always on the cusp of being too much.

Everything felt very tender, still. My close friends were still handling me gently but the news had reached those I'd not told. The break-up was, to them, now distant enough to be something of a curiosity, an opportunity for others to impart unsought advice and anecdote. When I told them that Josh and I were still on good terms, they'd scoff, tell me that would change, and my head would swim with trying to work out if I was naive or if I, too, would wind up as vitriolic as them. People would openly marvel at the mess of my situation – namely, the fact we shared a flat between us that we couldn't yet face selling or renting out, and thus we became a cautionary tale – and would find little to offer other than a kind of flabbergasted, half-baked consolation. It only made me make my guard more impenetrable, shrug off the doubt, crack a joke. I was getting good at it, layering up these brittle shells. I no longer existed as the long-term girlfriend, but felt that being somebody recently heartbroken or wilfully independent didn't work either. I had to work out who I was, underneath what was left from the mess of it all and the intervening years, and what I would do with myself.

I could up and leave, or find a flatshare, or give my heart to any fool and have it broken over and over again. Become a nun, change my career or abandon it completely and travel the world. I could move home. I could join the circus. This was what I was told, along with a lot of other people my age: work hard enough and you can do what you want. Who, though, imagined all those options could be daunting, that there would be some more acceptable than others?

Now I was rootless too, my plans vague and options so piecemeal they might as well have been those late summer leaves crisping and floating to the ground. I had tried to impose a little order on the fact that I had to move out – it was Josh's turn to return to the flat. August had been compartmentalised into different weekends, with new sleeping quarters for each one. To rent somewhere for longer than a couple of weeks seemed beyond me at that point, and I was still leaning heavily on the need to escape at the weekends, which felt so ripe with memory they risked rot.

Being the youngest of three in a loud house had meant I was always a bit scared of my own company. I would grow bored and restless if not occupied, struggle to get comfortable and scroll endlessly through Facebook, Twitter and Instagram, sinking into other people's friend-filled feeds like acid, even though it etched through my self-esteem and happiness. A lot of the time I felt that to be socialising was to be successful, that it was a failure not to have friends or not to be with them, doing fun-looking things, as often as physically possible. During those days at my parents' I had panicked and stacked up festival tickets and train tickets in equal measure, but now those weeks had rolled around and the sheer upheaval of it all felt so heavy. I'd overcommitted, over-planned. I was exhausted, desperate for some time to sleep and read and be a bit normal. To take my time over getting somewhere, without constantly scrabbling for the next bus or train, sweat gathering under my fringe.

I chose to push through it instead. The hangovers never really came. Someone quipped that it was party karma, a consolation prize for having my heart broken. But it felt more as if I existed on a single plane that simply pitched either higher or lower, depending on events. Days felt like the street confetti that graced the pavements in the summer, fallen petals from hanging baskets and smashed pint glasses, glittering and lost and sharp.

Between the nights spent under canvas, sleeping through the distant, spine-shaking thud of a festival heartbeat, I spent the first couple of weeks of August on the eighth floor of a high-rise in Battersea, a tiny spare room in the home of a friend's newish boyfriend. Josh was there when I left the flat for it. He packed me into an overly large cab and it was quite the strangest, and in some ways kindest, goodbye we had ever shared. I made myself appear upbeat and peppy, as if this were a fun summer excursion and not, in fact, the beginning of a tedious new reality. Inside I was cracking. The bed took up most of the room, and I'd roll over onto my stomach, look from the pillow to the train tracks that lay outside the window.

Summer was now into its seventh week. The mornings sagged, thick with a residue left by the sweat of millions trying to go about life in a city not made for high temperatures. Already, the sun was starting to languish. I'd stand in this new, unknown kitchen – bare feet on a sticky floor – and watch it, a cosmic yolk suspended above a docile city. A sign that the mornings were getting shorter. The whole thing felt stifling; being at home without Josh was punishing in its own way, with its constant reminders that he was not around. But to be somewhere new and different entirely, somewhere purely for shelter, was worse. The small acts of daily life – getting washed and dressed each morning, packing a bag and leaving the house – were all rendered difficult and alien. I had somehow lost the ability to function around the kind and friendly people who were, I suppose, my temporary flatmates. I found the trite dance of small talk so stupid; it seemed so unnecessary – I'd be gone in a matter of days and these people didn't need new friends. I'd always found cooking dinner a way to unwind and relax but could muster no motivation to do so, so accidentally melted ready meal trays in the oven instead. The people whose flat I was sharing retreated to their bedrooms in the middle of the evening, and so I occupied the living room, watching Netflix and staring at a screen. I imagine they were doing the same upstairs. Perhaps we were all feeling as separated from each other. I ached to connect with something, with someone; I went on nights out with odd combinations of people I barely knew, just because they were there and willing and I thought that brushing against other sweaty bodies in nightclubs would offer community. The old university friend would text occasionally, dropping breadcrumbs about words I'd published or sights he'd seen abroad, and I'd take to them like fish hooks, only to be left gasping for air. When I next saw him, I realised I'd constructed a vapid, impossible fantasy; there was no good to be found here, only fragile memory.

The cabin fever wasn't helped by the lack of outdoor space. The block of flats sat on a main artery that ploughed through south London, and you could see the city sprawled out below the sun-beaten glass. But I'd look out to admire the well-tended beds of the paved gardens dozens of feet below. Planting showed them to be beloved spaces: hostas, abundant and high enough to avoid the appetites of snails; jumbles of geraniums in full flower; neatly trimmed shrubs and masses of jasmine, a dose of sweet scent to cut through the stench of summer in the city. The tumult and swell of me just trying to cope pushed against those warm windows, and found only feeble puffs of grubby air. During the weeks, the city seemed to be made only of boxes: I'd wake up in one; take another down the train tracks to work, where I was lucky to sit by a window that was too high, too large, to open. If I looked out, there was just more glass, segmented into boxes, filled with rows of desks like mine. I couldn't calculate the years of my life I'd spent penned into a work station that I'd smothered in books and notecards as some semblance of identity. It seemed like such a waste.

I tried to escape it on the bike. Cities put such confines in place. Pen us in by ambition, paving slabs, traffic. A red light holds

like a regression to my student years; it was difficult to go and not be reminded that I was no longer nineteen and easily geed up by sickly cocktails. Unsure of what else to do, we headed back to the club that had been my favourite when I studied there. On a given Tuesday in term-time, nearly a decade earlier, I could walk in there alone and feel like I was at a house party as I knew so many people. But now, in mid August, we were in the throes of the summer holidays. The dance floor was deserted. The three of us being on it made it seem even more empty, a pale and mechanical imitation of the freedom and joy I had at a time of life when none of this mess existed, when heartbreak had been simpler. I felt faintly ridiculous for thinking I could recreate it; another woman stood where my younger self had been. I posted the photograph on the internet for Josh to see, waited for a smattering of likes to roll in with shallow gratification.

Doing new things was better. We walked the dog in Jesmond Dene. It's been a public park since 1866, and before that it was the extravagant back garden of the townhouse of Lord Armstrong, a local-boy-done-good who got rich from inventing weapons and turned into one of those Victorian philanthropists who spent his money on fantastical creations. When Armstrong and his wife (who had more input in the gardens than some fine ladies) got the land, the glacial valley would have been a scraggly mess of gorse and brambles, with a few indigenous trees. By the time the Dene was given to the public, it was a fertile fiction, teeming with waterfalls, mills, banqueting halls and an imposing iron bridge.

As a student, I used to puff my way across this bridge several times a week, leaving my home in Jesmond for others in Heaton, but the Dene remained largely unbothered by me. I had only properly explored it since graduating – my final hours of my time in the city were spent in the Dene's quarry, dancing by candlelight at a teeming party that wasn't supposed to be happening. For me, the Dene will always be the smell of wild garlic growing as exam season approaches, a combination of nature-based escapism and toil that, funnily, harks back to its original intent. Armstrong wanted to give the people of Newcastle, that soot-smothered, hard-working port town, a space to avoid the grind. Centuries later, it still met that brief. Nestled inside the valley, the Old Mill was roofless, but blooming with alliums in their last gasp, sweet peas and heucheras. The whole thing was covered in metal netting, but somebody must have gone in there and planted them. Some of the yellow poppies had started to go to seed, and I pocketed their drying heads.

The further I got away, the stronger the beat of London became. Even with its strictures, its frustrations, I felt tugged back there. And anyway, August was always a bad month to try to take off from the office; the desk hollowed out with too many school holidays and arts festivals. With home undefined and wonky I had begun to think of it as a space safe if only because of its lack of demands on me, somewhere where I could pretend that everything was normal and where I knew what role I was playing. I let my work, which was usually fast-paced and creative, meander, become a balm out of sheer boredom alone. I'd mostly dwell in drawing up spreadsheets and making sure the work-experience kids had the right amount of stuff to do. Before the break-up, I'd feverishly applied for jobs and ambitious fellowships in other countries, hankering for a change I couldn't define. Now, though, I leant into the sparseness of it, idly contemplated the freedom of freelancing while tapping out Google-focused pieces about where you could camp at Reading and Leeds festivals. It offered an anodyne counter to the chaos. I parked my suitcase under my desk and carried on as if everything was normal.

Being stripped of the solace of the balcony set me, without fanfare, on a mission I wouldn't realise for some time: to find the restorative benefits of green space elsewhere, beyond the remits of wherever I called home. The nature of the countryside may have been the one I was raised against but I'd only learned to seek it out in the city. If anything, I find greater kinship with the things that grow against the metal and concrete, where London's nature lies. The unruly life that defies planning law and pavement to chart the passage of time.

That was what we found in Hannah's garden. She and my brother-in-law had barely lived in it. They'd moved some weeks earlier, little more than a collection of days, really, into a white-washed three-bedroomed promise of their future. And that future was nudging firmly at the present: my sister was into her third trimester, her bump almost comically attached to her tiny frame. A soft-hard lump of baby; I remember her hand showing mine where to touch. 'That's its bum!' she'd say, and I could feel it, the little unborn rump a world away inside the same room, on the same sofa.

I slept in what would become the baby's room, folded up against the wall where his crib would soon lie. But there was space enough and I piled up my clothes – too many of them and of a strange assortment – around me, like a nest. It was on the side of the house where the light fell first during the day, and I was woken one Saturday to the sound of scrabbling and of earth on metal. Looking out, I saw Hannah bent over, legs wide, attacking the hard, parched summer ground with a fork.

I went downstairs, chastising her for doing such work so pregnant, only to be hushed – it was just eight a.m., they were new in the street and she didn't want to disturb the neighbours. Hannah had been unearthing weeds, ones that had long lain unchecked and had, meanwhile, fleshed out the bare soil with riotous foliage in lieu of the more favoured grass. Here lay round, scalloped leaves that belied the vigorous growth below. She was separating them from the ground and pulling, belly as bolster, trying to grasp enough to unearth roots that had become as hoary and thick as old carrots.

These were taproots, the long, pointed, efficient things that push out of a seed and deep into the soil to find and deliver water and nutrients to the growth above. Gently pull on a seedling or a small plant and, if it's the type to have one, the taproot will be the most impressive one of a bundle underneath. A surefire way to hamper a little plant's potential is to sever this – it won't grow as tall or as strong if the taproot has been damaged. But if they remain intact, taproots will be the most determined of the lot. While others will form a collaborative mesh, constructed to ground a plant, taproots are insistent from the minute they emerge from the hard, brown skin of a seed. They will go far if need be; in the Kalahari Desert, roots have been found sixty-eight metres below earth.

Whatever the plant, as long as it grows in soil the roots are after the same things: water, nutrients, oxygen. The oxygen is for their cells – leaf cells generate it as a waste product from converting the carbon dioxide in the air – but the roots must take it from the ground around them. White roots are ones that are breathing well: I've rarely bought a plant without surreptitiously upending it from its pot to check the root ball that lies underneath. How closely does that webbed mass fit the shape of its plastic cage? How wet and dark is the mass of it? Bright white roots; damp, slightly crumbly soil – now there is a plant that is ready to be bought, one with a good chance of survival. Too much water drowns out that air: the root cells can't breathe, the roots turn black and rot, and the plant eventually dies. Not enough room is almost as bad: roots will grow to fill the space. A root-bound plant will push at its boundaries, come to take on the shape of the container it inhabits, even pushing out at the holes in the bottom where water is supposed to flow. They need more – more nutrients from more soil; more space to expand – in order to reach their full potential.

The roots in Hannah's garden did not come easy. They were frustratingly brittle, and would snap in our slipping hands, the bright white (healthy, oxygenated) fissures revealing themselves, glinting in that south-east London clay with a sense of undefeated satisfaction. But we were as stubborn; we always had been. She is four and a half years older than me, and had longed for me to be a girl. I was – biologically, at least – but in childhood sentiment something far more muddily hewn; less a tomboy than a person motivated by supposedly boyish things: my brother's Britpop CDs, hand-me-down Lego, an obsession with navy, an abhorrence of pink.

The three of us had always come together through practical, time-occupying tasks. In the late Nineties, we took over a small plot of earth across from the house and set about turning it into a vegetable garden. We planted seeds, laid potatoes in the earth, set lettuce plug plants (there is a greater initial triumph in raising from seed, but those that have already been raised for a few weeks are sturdier, less susceptible plants. Most gardeners use a combination of the two) in rows. My mother was supportive, if sceptical. She said we wouldn't tend to the patch as the months warmed up. That we wouldn't weed, or thin out the carrots – getting rid of the weaklings to let those stronger orange roots (yes, taproots) fatten. That we would lose interest as the slugs gained theirs. This would be the most expensive way to gain a meagre portion of greens.

She was right, of course. All of these things happened: we had a collective age of thirty-four, and mere iotas of patience. We also didn't really like vegetables that much. But the act of working on something, no matter how ill-fated? That was something we had long done: building snowmen; creating cardboard palaces; burying our dad under sand on the beach. Here, decades later, Hannah and I were slipping back into what we knew. For the first time in so many weeks of every single action arriving with a question mark, to spend those post-dawn hours quietly weeding just made sense. There was little discussion over it, instead I fell into its rhythm: spade, scrabble, tug, spade, scrabble, tug. The gratification in removing – carefully, firmly, slowly – an enormous, unwanted thug of a root from the soil beneath our knees was visceral; the pleasure of it squirmed in my gut. We chucked them in a bucket, the thud of it was gratifying.

I was conscious that this was one of the last instances of us doing something – spontaneously getting on with it, as we would later spontaneously tour her garden – while we existed as just sisters, while I was still the youngest. Within weeks, the baby would be here. Another version of her, one with biology closer to hers than mine. Something she longed for, again; only inside her own body. And this time, if she had a preference on the sex, she wasn't telling us.

Gardeners pause a bit in August. The splendour of May has long passed; the riotous growth of June and July slows a little; the cut greens have come, and come again. And while seeds can be sown for autumn and winter colour, many experience the eighth month of the year as one of deadheading and avoiding the devastation of drought. In those first few weeks between new house and new baby, Hannah poured that swell of burgeoning maternal feeling into the new garden. She watered relentlessly. Her desire to hydrate her plants bordered on obsession. I'd find her doing it in the mornings, or after work – the same time my mother always had, because it was a pleasant way to spend the minutes between putting the dinner on and pouring a sherry, and it's the time of day when less water evaporates and more soaks into that parched earth.

Friends who have had children have said they found themselves needing to garden while their babies were still growing inside them. It seems to be an unspoken thing, perhaps because it is such a subconscious one, buried within the functioning matter of our bodies. One even compared the two: 'parts dull, parts horrific, parts a huge worry but very worth it in the end'. Another said she only stopped gardening during her pregnancy because of the winter, but would have gone crazy had she halted any earlier. One woman, pregnant with her third child, gardened throughout her pregnancies. 'Garden started as nothing three years ago,' she told me, 'now it is thriving.'

While we are yet to shift to a time when people ask if the child-free would like to bear children, rather than when, the expectation around women's bodies has softened in comparison to those put upon our mothers'. My maternal feeling has always been a sluggish, idle thing and I don't know if that will ever change. Yet others speak of that urge as if it were a heart-wrenching metronome. I remain grateful that we have more choice to do what we want in either case. Our generation are taking longer to grow up, to buy houses and put families in them. In my late twenties, I felt fertility creep up around me as friends fell happily pregnant and I remained ignorant of whatever was lingering in the water. But perhaps that need to nurture started to present itself in a different way: specifically, in the seeds I found myself willing to germinate in pots of soil on windowsills. There is much irreverence over the growing millennial fascination with plants – that we are too immature, too incapable, to cope with a puppy or a baby, so we fuss over expensive tropical houseplants instead. But such glib statements ignore a more primal need, one that has shown itself time and time again over the centuries: that if you take humans away from nature, put them in boxes in lands made of concrete and asphalt, sit them in front of screens and transport them in moving metal, hidden away from the open skies, they will seek greenery out. Our generation is not the first to turn to other living things for comfort – whether by procreating, or otherwise – and we shan't be the last. As I was witnessing, in myself and in others, when all felt tumultuous the rhythms of nature gradually became a siren call, offering a steadiness I couldn't grasp elsewhere.

In Hannah's absence, I was instructed to continue with her maniacal hydration routine: grass seed, a badly kept hand-me-down from my parents' shed, had been scattered as a last hope against the patches of the bare clay left behind by our root-upheaval. It was a dry August, and that ground stayed hard. I watered, tugging the hose (a hose! An outdoor tap! Such wonders of an adult-sized doll's house for her, with me stashed away in the back room on sororal favours) across to the two uneven beds on either side of the lawn. But I rescued and planted, too. There was a basil plant that was drenched daily, despite me advising Hannah otherwise. After every watering, I'd move it into the sun only to find it moved back the next day. Basil is native to India and enjoys a warm, dry climate. It needs sunshine and heat and, like rosemary and lavender, hates being left with wet roots overnight.

August is not the time to plant much, but pansies would certainly not be on that list: they like it wet and cool and they will withstand winds far more ferocious than their size. Their floppy-petalled heads turn to face the sun, but too much of it will kill them. I planted them nonetheless, those neat little cubes of soil, wrapped in white roots and wet between my fingers and thumb before I nestled them into new holes in the bed. They lasted well in spite of the heat, even spreading out of their patch and into other beds. I took the poppy seed heads from the Dene to the other end of the garden and ran a fingernail along their fluted ridges. Tiny, perfect black spheres landed in my palm with an imperceptible patter, and then onto the bare earth below.

Hannah aspired to grow buddleja. I'm not sure whether it was because she had seen it proliferating across the city's railways or because a huge bush of *Buddleja davidii* had grown in our childhood garden – next to the doomed vegetable patch, in fact. Those fat cones, composed of dozens of trumpet-shaped flowers, lured in butterflies like nothing else. Commas mostly, their tiger-patterned wings shining against the purple, but peacocks and small tortoiseshells too. A humble menagerie that accompanied school summer holidays. My dad would point it out: 'That's the Butterfly Bush,' and then I would too, endlessly, proud of my modicum of vague botanical knowledge. In full bloom it emitted a smell like the taste of honey hitting the roof of your mouth, a smell so rich it could almost turn sour.

With adulthood, summer happened earlier. The annual love affair between buddleja and London's railway stations would start in July and ravage its way through the city until autumn. It should be noted, at this point, that buddleja don't have tap roots. They're not plants to linger and settle in one spot. Rather, they spread a web of clinging, fibrous roots to claw as much nutrition as they can from the brutal landscapes that they inhabit. It is a plant made for making the most of a space – grow, flower, seed prolifically – as quickly as it can, in the knowledge that it could be easily uprooted at any time. It's a divisive plant as a result: gardeners must balance their appreciation of its aesthetics with scorn for the fact it is invasive.

By summer solstice – in London at least – the buddleja stretch their shoots to cover tracks and stations with tight, green cones of buds ripe with potential. They look skinny, but they're strong enough to withstand the gush of a passing train. Stubborn, too. Buddleja branches will nudge through the cracks in roofs and walls. At first, this will be just a curious stem. Then, by the high summer of August, the blooms will be heavy, bringing adventurous insects into the bustling chutes by which humans go about their day. The romance is in full swell, its infatuation paints itself over the station furniture in shades of lilac. As the month wears on, the bright violet of those trumpet-shaped flowers turns brown and dry as the plants divert their energy into the 3 million seeds each one can put out before it's done. Featherlight and designed for opportunism, these seeds are lifted and carried by the warming whoosh of air let out from underneath the speeding trains before rattling to a rest to germinate and grow.

That summer was no different. The weather shifted into the first proper summer's day and the buddleja left leafy curlicues in the corners of the Victorian glass above Denmark Hill Station. That plant is one of my favourite adventurous butterfly bushes; it looks so eccentric, bursting out of dirty glass, too distant to be maintained. Snow falls on its skeleton in the depths of midwinter, a ghost of the extinguished affair. It may never have planned to stay here, but it has on sheer tenacity, long after the seeds have been sown by means of a dozen trains an hour.

This frantic, smothering tryst has been happening for decades. It is not the train tracks that buddleja loves, but the lime in the ground left there by the mortar used when London shot up in spidery veins of streets of tall, narrow terraces. That lime mortar leached into the railways and the buildings around them, too. When the city's houses fell during the Blitz, the buddleja grew. So much so, in fact, it gained another moniker: The Bombsite Plant.

Untold human loss was caused by the bombs dropped in the dregs of the thirties and the beginning of the forties, but nature reclaimed the land in London. The churned up, abandoned ground made a surprisingly accommodating home for enterprising plants, which saw the lack of boundaries as opportunity for new beginnings. Bombweed, or rosebay willowherb (*Chamaenerion angustifolium*) was swift to flourish. It made good use of the railways after being introduced to a few of them by the Victorians, but it thrived in the light let in by the fallen buildings and monopolised the soil. Rosebay is called fireweed in America because it prospers in ground left charred and broken by forest fires, areas where life has been so destroyed only it can colonise what's left. By the mid Forties, 90 per cent of the bomb sites examined by botanist E.J. Salisbury contained rosebay. You can still find it now – not in gardens, for it is far too pernicious to be willingly introduced – but lining the railways in clouds of dark pink, its arrow-headed spires standing several feet tall over the land they have colonised. As roots give stability to the plant, so they give stability to the earth. Rosebay doesn't just grow once the flames have died down, it is actively introduced to American lands because its ferocious root system – a long, efficiently branched tangle that spreads far and quickly – can bind ruined soil together. From there, other things will grow.

And they did on the bombed sites, dozens of wildflowers and grasses that turned whole swathes of the Square Mile and east London into meadow. Until 1940, inner-city London was a smutty hole where few plants other than mosses and algae grew freely. But with space came life. Finally, those hopeful seeds carried on the undersoles of shoes, in someone's coat pocket or cigarette case, the ones that escaped horses' nosebags or dropped off the fur of a shaking dog, settled in the disruption of wartime London. Richard Sidney Richmond Fitter, better known as R.S.R. Fitter, was born in Streatham and grew up to become a wildflower expert. In 1945, he published *London's Natural History*, which charted the undulating battles waged between man and nature since before the Romans turned up. He also passed on his knowledge in person, by taking groups of teenagers on wildflower-hunting missions through the bomb-scarred city. Jane Lindsay was one of them, and she collected and pressed her findings – purple toadflax, pellitory of the wall and lucerne – in a journal that was later passed on to the Garden Museum. She found these flowers on the ruined land that would later contain the Barbican, but Lindsay just remembers it as 'a wild, open space, full of birds and wild flowers . . . and remnants of the old Roman walls'.

There was another buddleja that I followed keenly, a handful of stations away in Hither Green. I'm not sure when I clocked it first, it could have been months, or even years before. But I certainly took note that August, when I dragged my unnecessary luggage from SW11 to SE13 to pick up the keys of another bolthole, frustrated and sweaty with the faff of it, although the weight of the brave face was far heavier. That buddleja was a squat, sneaky thing, colonising the bricked-up gap between the faded wall and the mucky tilted windows that lay overhead. A flash of surprise against the corporately chosen pale mauve of Southeastern Trains. It dangled on the left with new oval leaves neatly spread down either side, like an Alexander Calder sculpture.

I admired its persistence and took pleasure in how much it liked the spot it had forged for itself. It had grabbed some prime real estate – bright, indirect light, good ventilation and, from somewhere in the roof, water. With each day, I watched it grow. After a couple of weeks, I stopped making that journey daily. But I still went back to Hannah's; our shared roots made her new, empty home a place of stability, offering a kind of familial structure. There were conversations over cups of tea, but also pasta eaten without picking over the pain I was in and mindless, comforting television viewing. Getting off at the station, I'd notice the difference. As the baby grew and Hannah's stomach spread, so the buddleja took up more space; now the stems were fuller, they reached the bottom of the glass. Those violet, grabbable cones hung mere inches above commuters' busy heads.

It offered a different means of understanding the passing time than my scrambled thoughts could offer. I was beginning to feel better; my happiness was returning. But it didn't grow regularly, on a daily basis, like this buddleja did. To turn up and see its changed shape and stature offered a greater sense of change than anything a calendar or clock could show. Dates, numbers, those

SEPTEMBER

I WAS TWENTY-ONE WHEN I left for New York. Of legal drinking age, but not really experienced enough to do much else there without learning about it along the way. My dad took me to Heathrow, pulled up at what in America is called the 'Kiss 'n' Fly' but over here is just known as 'drop off'. Watched as I lingered at the shining revolving doors of Departures just long enough, before telling him that if I couldn't check myself onto a plane, I wouldn't stand a chance of surviving once it had landed. He said goodbye like I was going out for the evening, with the kind of hug that was formal with love.

I was thrilled and scared in equal measure. Had a three-month internship at a now-defunct youth lifestyle magazine four days a week, but nowhere to live until a fortunate phone call came in while I was at the gate. I'd managed to get a place on a journalism masters course in London but somehow that felt even more risky, even more indulgent than packing off to another country with cash I'd scraped together from my student jobs. The fees were so high, rent was ludicrous, I hated the notion of being in such blatant competition with other people on the course for the same handful of opportunities. Newspapers had always run graduate schemes, but in the wake of the economic crash those had stopped. At that time in London, internships were little more than work-experience stints that went on too long. I'd been working for free in magazines and local newspapers during the holidays for the past few years; couldn't bear the thought of doing more of it just yet. Figured instead that it made as much sense to be in New York, where I could see some of the world and they seemed to know better what to do with us. It was a powerful combination: that sense that we could do what we wanted as long as we worked hard enough, combined with the increasing evidence that there just weren't enough jobs to go around. I let it stew, become a fierce appetite, the kind of thing that made me go through with something so fanciful it was the plot of several movies: move to New York, follow your dreams.

I soon learned that staying in the spare bedrooms of professional people was not the same as living with fellow students who wanted to be friends. That in America, the oven temperatures are in Fahrenheit, and in New York nobody really uses them anyway. Loneliness, proper loneliness, arrived in my life for the first time and resonated with the hollow depth of a distantly ringing bell, gently ricocheting through my twenties from then on.

From Friday morning to Sunday evening I'd explore the city, the parks and the delis and the museums where I donated a dollar to visit. I'd rattle on the train out to Coney Island and stuff myself on $4-worth of Chinatown dumplings. Nights were fuelled on the free, under-drunk bars of fashion parties and deals in dives, and it felt surreal in its otherness, the new glamour of it all. I'd rely on the breeze pushing through the platforms of Hewes Street and Marcy Avenue stations to blow away the treacly hangover left by PBR swigged from cans. The glossy Manhattan shaped by movies soon became something I only saw for a few precious seconds twice a day, heading in and out of the city from Brooklyn on the M train, light hitting the Chrysler Building. Instead, I formed my own, different New York.

I found a small cabal of other people, mostly students or graduates, mostly ex-pats, who had also reasoned that it would be expensive and difficult to become an adult, whatever we did, so we may as well do it the way we wanted. Comfort came from a couple of Irish girls who lived in a railroad apartment on Montrose Avenue and made condiment-slicked dinners that, along with the company, softened the side-swiping homesickness I'd never known to expect. We traipsed the city together, spent our fledgling office hours communicating fervidly on various instant messaging services about drinks and dates and sublet rates. Those messages still linger on the internet now. To look at them is to read the slang and the sentiment of different people. How quickly we fell in love with one another; how bold and brilliant we thought we all were. Years later, our lives would collide again in unexpected ways and within hours we would reflect, affectionately, at the relative children we had been when we met.

This New York was built from other bricks. Of the burn on the roof of your mouth from $1 pizza slices and the inability to take the G train without getting lost. It occupied a period of time as small and undulating as the circles I spun one night on a rooftop we'd snuck onto. The lights of the Brooklyn Bridge drew streaks across the teeming sky. I grew streetwise, discovering how to dodge the drug addicts who would shout late at night, thrusting guns made of fingers in your face, but never tough. It was naivety and sheer good fortune that kept me out of any real trouble.

Those weeks were like candyfloss, gritty and light, as crystalline as Manhattan sky. Like its inhabitants, New York air has a sharpness; the light rolls fresh off the Atlantic and collides with those heavy blocks and gleaming spires, whereupon it splinters and exposes, leaving crisp shadows and little room to hide. That light would shatter down those island-spanning avenues and carve the sky into rectangles, fill them with pink clouds. Down on the ground, the New Yorkers went about their business and I looked up. The height of it all never grew old. When darkness falls in those last days of summer it catches you by surprise.

Manhattan's architecture buttressed against that light like commuters crammed into a morning subway cart. But in the parks a dance was unfolding, of sunlight on leaf held together by flurrying, late-summer winds. It took me some time to get up to Central Park; I spent those first weeks in the city learning the tree-lined avenues of Clinton Hill in Brooklyn and lazing on the expansive lawns of its Prospect Park. Williamsburg's McCarren Park was the bit of green space closest to that first room I sublet, but it felt sad and grubby; I went a few blocks over to East River Park instead, stared across at the skyscrapers.

As the sticky, oppressive palm of summer lifted, the colour crept in and I made my way up the other island. I would span in overwhelm. They'd been in town a bit longer, and took me in. She was working at a publisher's in Chelsea, underneath the High Line. Little less than a mile and a third of the former West Side Line, a railway that took cows to what remains known as the Meatpacking District, it withstood New York's rampant pace of building and cries for its demolition for thirty years. Instead, it was turned into one of the greatest parks in the world.

After days of trying to settle the mythic, cinematic New York with the heat and trash of the real city, to stand on the High Line for the first time was to occupy that blurry space between the two. This was when the High Line was still an infant, before it became the fourteenth-most recommended thing to do in New York City on TripAdvisor. A scrub of land that was newly opened but without that reflective polish, as if it had been born old and knowing. It was a space suspended above the relentless pace of the sidewalk but beneath the calm that comes from gazing out of the city's high-rises. A giant found object made for one intention but left unloved until another came along. The High Line was a man-made space that man forgot, and by the time it was reclaimed it had become somehow other; an ethereal, shape-shifting thing rooted in both nature and machine. The light and wind rippled through it, preserving its image as quickly as it whisked it away, a perforation of Chelsea's landscape that was at once undeniably a home for nature and yet a deeply unnatural being. Caught between warehouses that had become artists' studios and lofty living rooms, the High Line was both an artery and an escape – something vital and yet strangely placed. In the city known for turning dreams into reality, the High Line transformed

Manhattan into a waking dreamscape – something that seemed to exist beyond its very being.

In the thirty years after the trains stopped running along that bit of the New York Central Railroad, people didn't go on the tracks and wilderness grew up instead. During the bright new dawn of the millennium, photographer Joel Sternfeld was invited up to the former railway there by Joshua David and Robert Hammond, two men who lived nearby and dreamed that it could one day be a garden. Sternfeld was astounded by what he saw. 'Suddenly you're in another world,' he said a few years later in *High Line Stories*, a short documentary film. 'There were wildflowers, there was broken glass, there were birds, there were mourning doves.'

Sternfeld spent the next twelve months accessing the High Line whenever he wanted, while the vast majority of New Yorkers still had no idea of the world that had grown above their heads. With time, he realised that its raw beauty was best shown by looking straight ahead, down those narrow tracks into the cityscapes and water beyond. His photographs, from a series called Walking the High Line, capture elevated land that has gone to seed: rusted tracks disappear into a mass of browning grasses swaying above the ornate tops of iron streetlamps; the red-bricked buildings and chimneys of midtown Manhattan are separated by an early summer flush of willowherb foliage and sprawling mounds of morning glory. Delicate yellow wildflowers chase a curve in the line in spring, while young trees bridge the narrow gap between warehouses that would become boutique apartments within the decade. Sternfeld went up there in the winter, too, and set up his camera in the snow to show the brave scaffolding earlier seasons had left behind. Four years later, a study published in the *Journal of the Torrey Botanical Society* found that 161 plant species had been belligerent enough to seize the fertile land left behind by man. Eighty-two were indigenous, but the other seventy-nine had come to the city from elsewhere and managed to put down roots, like the millions of New Yorkers walking beneath them.

There were some who wanted this space to remain as it had been created, grown in the absence of humans in one of the most densely populated cities on the planet. David and Hammond formed Friends of the High Line to galvanise their intentions to transform the space. In its infancy, Friends of the High Line consulted locals on what they thought should be done with these eight acres of empty, beautiful space. Hammond received a card that said: 'The High Line should be preserved, untouched, as a wilderness area. No doubt you will ruin it. So it goes.' Hammond kept the card posted above his desk because it encapsulated his biggest fear, 'that we couldn't capture that naturalistic beauty in its wild state. That we would ruin it.'

Rather than preserve or radically reinvent what nature had created there, the Friends harnessed the crux of the High Line's spirit – the ephemerality of what grew there as much as its persistence – and channelled that into its planting. The planting there was so present, enough to transport you up and away from the stress of the city, but also somehow animalistic. While the rest of Chelsea was neatly stuffed with foliage elegantly tumbling from street planters and window boxes, the High Line seemed to breathe; it was alive with a tension and an ownership of the space that felt defiant of human involvement. The plants held the wind and seasons and lived within them; frills and fuss were sacrificed for structure and scent. This was a meadow in a box, suspended in a secret spare rib of the city. I'd never been anywhere like it.

What's so striking about looking back at Sternfeld's photographs is how much they resemble the High Line of 2010 and the far more developed High Line that followed, even though the park now acts as an art gallery and a tourist attraction to millions. Those who transformed it looked at the flora that was there already and used that as the drawing board. Dutch designer Piet Oudolf was brought in to decide which plants would go where; in doing so, as Sternfeld put it, he achieved something 'that didn't look very likely at all at the start'.

Oudolf's gardens exist out of fashion and time because they are created with less transitory notions in mind. Not of aesthetics as much as feeling, space and beauty. 'A garden is also a promise,' he said in *Five Seasons: The Gardens of Piet Oudolf*. 'It doesn't have to be there, you're looking for what will be there.' Oudolf is a plant fanatic, a purveyor of arresting beauty who actively encourages the birds and the bees. A man who resisted the counter-culture movement of the Sixties but nevertheless pioneered the benefits of gardening naturally with 'no pesticides, no artificial fertilisers and no army of gardeners to keep the plants alive'. His approach is radical. Perhaps it's because he's one of those gardeners who was not brought up around plants. Rather, like me, he found them in his mid twenties and saw them as a means to escape the career intended for him: working in the restaurant his family owned. He said he wanted to 'do something more', a sentiment that resounded loudly with what I was feeling.

Along with plants people such as Beth Chatto, Oudolf changed not only the way we look at gardens, but the plants we grow in them. His students find him inspiring and infuriating in equal measure: he has no singular process, and frequently makes his planting plans public – once used, he has no need of them, preferring to come up with new ideas instead. He's now considered the most prominent member of the New Perennials movement, a shift that happened across Germany and Holland in the Eighties and that made gardens into something challenging and alluring in equal measure. The basics of it are simple: matching plants to the situations in which they will thrive best, and using ones that will emerge, flower, seed and die, then repeat the process a year later – these plants are known as perennials. But perennials also require patience. Annuals guarantee near-instant results: a pop of colour arriving in a bed from the garden centre. When they fade, they are whipped out and replaced cheaply and cheerfully. Perennials, meanwhile, demand that the gardener appreciates a plant during all of its life stages, even during those when they are invisible, lurking beneath the soil. In exchange, they offer structure and surprise as the year turns.

In Oudolf's gardens, each season brings a different drama. The appearance of new shoots through snow in late winter paves way for a soft riot of spring colour. Summer's showiness leads to the lazy *petite mort* of autumn. Inevitably, the garden becomes a grand closet of skeletons. Then, in the low light of misty, late winter days, the whole thing is cut back and left shredded on the ground to feed the earth and the patient life below. What comes in its wake can be predicted, but will always contain a few surprises. Change is integral to these gardens' beauty, as it is to so many things.

In its infancy, many considered this stubborn refusal to tidy up old and lifeless growth a shocking thing in a garden. Oudolf worked closely with another Dutch gardener called Henk Gerritsen. Together, they wrote *Droomplanten* – or *Dream Plants* – a compendium of the 1,200 perennial plants they promised would make for an easy and beautiful garden at all stages of their life cycle. As Oudolf would say many years later in *Five Seasons*, a documentary about his practice, 'Death is the garden as well.' And it was Gerritsen who taught him this. 'We discovered that plants were good even when they were not flowering,' Oudolf has said. 'He pointed this out to me a hundred times. We looked at plants at times other than their prime time.' For Gerritsen, who lived with HIV until his death in 2009, and had lost his partner Anton some fifteen years before, appreciating the beauty of lifeless plants took on an even greater significance. 'People used to be so afraid of death in the garden,' he once told writer Noel Kingsbury.

> Every yellow leaf was an imperfection, and had to be taken out . . . but now a whole generation has known death, so we do not ban it from the garden anymore.

To grow in this way is to understand and accept that the end of something is necessary to foster renewal. That life needs its endings as much as its beginnings.

Having said that, the High Line is a garden built very actively in life. As field operations manager James Corner hoped in 2004, five years before it opened, it was always intended to be an area for:

> Hundreds if not thousands of people being able to walk up and down this thing and still celebrate something about the energy of plant life emerging through a hard surface.

In the end, they got millions. The visitors are something enjoyed by Hammond and David too. 'It's better with people,' Hammond wrote in 2016. 'The people are as important as the perennials.'

The High Line's planting, too, is a moveable feast. Oudolf's garden theory has never been static. He has said he builds them 'out of different experiences, different concepts' and when these change, the plants do, too. The freedom Oudolf gave to plants was one of the main reasons he was brought onto the project. When you're up there, among the swaying grasses and sheltering trees, it's very easy to forget that the High Line is actually one enormous container garden, a supersized balcony opened to the public. Even trees as large as bur oaks are rooted in no more than eighteen inches of soil. The plants are looked at and listened to; those that do not thrive are not continually, stubbornly replanted, but allowed to fade out. Siberian catmint – Latin name *Nepeta sibirica* – is a pretty, rangy plant that throws up tall spires of purple trumpets and has a web of roots beneath the soil. After a first dramatic flush for a couple of years, Siberian catmint struggled on the High Line, and so it was granted permission to dwindle while better-suited plants, such as hardy geraniums, took advantage of the room to spread.

'The one tool I can't be without are my eyes,' Oudolf told the *Wall Street Journal* shortly after the High Line opened. 'Sometimes you need a spade, sometimes pruners, but when you are gardening you really have to look.' And you can tell that the High Line is a space that is examined – not for imperfection or fault, but to see how things are working, if the tension is as it should be. This is a space that exists as an agreement to experiment, to suck and see. It's not about making good, but making sense of something challenging. To find potential in those gristly, unforgiving boundaries of the city and not only make do, but make better, provocative and different.

•

Every summer after, on the cusp between August and September, I'd dream about New York. The visions were shape-shifting things, varied in colour and cliché, but always of the same place. Some would be of walk-ups and the smart green porches that cover the entrances to Upper West Side apartment blocks. Others would plonk me on the sidewalks of Brooklyn's sprawling avenues, or be filled with the cobbled streets of SoHo and the rancid-delicious smell of the kebab stall by the southern exit to Broadway–Lafayette Street Subway. This febrile Frankenstein city would creep into my sleep for several nights on end.

I was always dissatisfied when I woke up, frustrated with the greying late summer mornings I'd find outside windows in Peckham, Hackney and Camberwell, feeling trapped against the expectations of a city I'd moved to for work rather than out of desire. That to be here meant subscribing to a life pushed onto me – long hours and climbing ladders and partying hard before settling down – that I had never had the chance to wonder if I wanted.

I think the dreams came then because I've always felt the shift between summer and autumn keenly. I was born on the autumn equinox, September 21. Later than expected, but fitting: it made me the third in a set. My brother was the first, turning up four days before Christmas on the winter solstice, and my sister appeared on the spring equinox. I think there was a feeling that there should have been a fourth child, born with the onset of summer, for the purposes of completion. But some decades later, two perfect, identical twin girls arrived into my brother's world on June 22. A neat enough symmetry.

I've always found great satisfaction in my equinox birthday. The fact that a new year of my life was starting with the shifting of the seasons is pleasing, as if a giant clock somewhere had moved its hands into a new arrangement with steady, clunking satisfaction. Autumn was a season not just of new pencil cases and falling leaves but one that, like fattening apples, ripened with the potential of change. I felt it more keenly than the thud of winter, the tentative awakening of spring or the oft-surprising blast of summer. I would relish those rare mornings in late August, when, if you got up early enough, there would be a hint of the renewed chill in the air: a harbinger of September.

This is when the burning begins. Summer dies out in a fiery blast, with the last gasp of surprisingly warm afternoons, picnics and al fresco drinking clutched at, the last ones there will be until the spring. Autumn is nature's beautiful death, the leaves set ablaze before carpeting the ground, letting in winter's ravages so that life can gather strength below the earth. I didn't crave demise but I needed to burn and brightly, to work out what needed to be shed so I could make room for the contemplation of winter ahead, so I could allow new life in the spring.

As long as I was coddled in education I was kept strongly bound to the new starts of that month, although I always got vaguely annoyed when my university lectures wouldn't kick off until October. But upon graduation that institutional calendar slid away. One was left to one's own devices, to chart progressions as one wished and as they occurred. I couldn't remove that persistent feeling of autumnal change nor did I want to. From early adulthood, September would arrive with a kind of niggling urgency that needed to be attended to. It was more than the businesslike arrival of the end of summer, when colleagues would return from their holidays, tanned and slightly resentful, or the endless commands from fashion magazines to buy a new coat. I was quietly desperate for something new.

But that seasonal shift became harder to recognise in the city. The signs have to be properly hunted out. The buddleja shrivels and darkens; stray rosehips swell in scrubby gardens and the communal spaces of housing estates. Suddenly Japanese anemones, those prettily persistent pink and white flowers, begin to bloom above brick walls and through iron gates and last until the first frosts. Blackberries darken on the stray branches that wind themselves over fences and by roadsides. The leaves on the trees won't turn for weeks yet but there will be more of them lying brittle on park lawns. Under the glaring sun, bouncing off the warm urban brick, people will still be in shorts and vest tops; the Tube will still be clammy.

It wasn't so much that summer lingered but that, as I moved through my twenties, it became easier to ignore the habits of nature that crept in around London's edges. My life became increasingly severed from those new-start rituals: house moves turned into a rudimental shuffling of convenience and cheap rent, my relationship progressed with a solidity more certain than I sometimes felt capable of. I would often look at my friends, who were struggling to date and struggling to work, with silent envy. It was easy to feel that I had been living my life too quickly, making a home before I was meant to and encountering the wrong kind of adulthood's difficult parts too soon. That to settle, among our generation, was a failure: the adventurous became digital nomads or chased their ambitions into reality, had dozens of flings with all sorts of people before ripping it all up, starting again. I was fiendishly lucky but it didn't stop me from hankering to know where else life was living, and how. Where is your new start, September would hiss at me, what have you done in the past year?

That girl who moved to New York never really thought any of this would happen: the nice flat and fashionable boyfriend and, especially, the writing gig at a national. And yet I'd turned into a young woman who struggled to find satisfaction in much of it. I'd lay out each little challenge: the byline photo, the front-page story, the feature puff and the viral hit. When I landed them – often after months of scrabbling – no joy arrived. Instead, I'd move on to the next one for another hollow victory. Each became a new bar to jump, like the exams we used to take several times a year. We'd been shown that to practise enough would result in achievement, but never taught where to find the satisfaction of getting something done beyond the sake of it. Everything held significance and show. There was always more to be achieved, always someone else doing better, who would work longer hours than you at the job you were lucky to have.

At first, I'd listen to the dreams, take them wholly literally as a message telling me to abandon the life I had in London and return to New York. One August, a year after we had started dating, my frustrations with the life Josh and I were building bubbled up into an explosive row on the banks of the Thames. I felt confined. By him, by the strictures of what having a serious relationship at twenty-three meant and said and felt like. By the notion that I should be exploring as much as I should be staying put and making a path for myself. I was stultified by the lack of change; I felt cramped, somehow, and probably because it meant that nothing was growing. I couldn't expand what was in my life because I hadn't found the room to change – and not my relationship, or my home, or even my career, but to clock the shift I needed to make within myself. To have the bravery to shrug off the notion of what I should have been doing, what I should have been happy with, and start looking for the things that actually made me feel something instead.

Ultimately, though, it seemed a mistake to throw it all away when it looked so good in theory. With each passing year I got better at ignoring the dreams that coloured my sleep. I'd toy less with the notion of leaving, although some things, such as looking down Kingsland Road at the City during the hazy light of early summer, would tug me back to Brooklyn without warning. Amid the late nights and long, lost conversations of my twenties bathed in cheap wine and chaotic social structures, a kind of empty satisfaction crept in by stealth with nice furniture and expensive fruit juice and the sense of belonging borne of living in love. I smothered that need for change in hollow gratification, a quiet muffling of making do. For years the fires dampened, those fierce colours stayed the same evergreen.

But that September, the one after the break-up, was different. There were no dreams. Six years after they started, those nocturnal demands for change stopped. I realised it while watching one of those chewy, late-summer sunrises on the train to work. Seeing that crisp air hit the stilled cranes that were propping up Nine Elms took me back to my Brooklyn–Manhattan commute. Until then, I hadn't thought of it – awake or otherwise – for months.

Those subconscious cravings for adventure, of uncertain swell, were no longer necessary. I had been granted all the unpredictability I didn't know I had asked for. But beyond that, I had been forced to examine my own needs and desires, rely on my unstable impulses to forge a life I wanted. I was burning. Beginning to let the remnants go in order to gather strength and return, with new growth, in the spring. Without really meaning to, I was somehow bidding farewell to the raw heat of the summer that had whittled itself away while I was fully conscious, rather than hankering for something else in the twilight of my sleep.

•

with love and productivity, snatching some lavender branches from one of the bushes on the estate and knifing them into cuttings, carefully placing them around the edges of a pot of gritty sand, with the hope that they'd root to grow by themselves next year. (If you put those newly severed stems by the pot's edges, right up against the plastic, the roots will hit the base when they develop – it gives them a better chance of entwining with one another, to create a supportive mass. Shove them in the middle and you can end up with roots plummeting right down the hole, too adventurous for their own good.)

The sedum was reaching its prime. I'd picked up this one a year or two before, an impulse purchase at the wrong time of year – mid autumn, when the plant was at the peak of blooming and wouldn't look anywhere near as good for another year – and, taking little effort to research what it needed, plonked it in a dark corner of the balcony, between two walls. Stonecrops – fancy name *Hylotelephium*, known to me as Sedum Autumn Joy – are reliable, life-giving plants if granted the slightest of chances, namely, a fair bit of sunshine. The one I had rapidly abused went a bit soggy and eventually died back. But in mid spring I noticed chubby little leaves appearing from the remnants of those broken branches and planted it up in a large zinc tub, which was precariously wedged between the iron railings and the wall ledge on the left-hand edge of the balcony. Had it fallen off, it would have killed someone, but it wasn't going anywhere. Exposed to the languishing sunsets, by midsummer strong burly sedum branches had covered that western corner. September rendered it a riot of showy, flat-mushroomed heads with flowers dwarfed by the bees that drank from them. A late summer luxury, the plant was relishing the moisture that had been left in the soil thanks to the near-non-existent drainage in the zinc tub. It was happy enough with the exposure, absorbing the light and admiring the view, managing to withstand even the belligerent gusts with the lot it had been given and bloom anyway.

I decided to give the flat new meaning, filling it not with loss but with a different kind of love. I'd cook simple, comforting meals for my friends, the kind of thing that doesn't distract from talking while you make it. We'd squeeze onto the balcony and watch the sun dip as we spoke. Sometimes I'd wake up and find them on the other side of the bed. Other times, they'd leave nothing but evidence of gentle signs of care: plates stacked in the sink, my shoes neatly lined up by the door. I could tell I was starting to function better by the slow creep of normal things into my life; in doing laundry and eating a proper meal, all because it had been a while since any of that had happened. The last time I'd slept in the flat had been July, when I had found the prospect so unpleasant I'd rarely done it sober.

It was less daunting now, but things still felt so quiet once everyone had gone home and the washing up had been done. When others weren't around, the loneliness still bit as evenings stretched out into unwanted languor. Most days were still filled with a dull aching that I felt most keenly in the mornings, which were quiet, and the increasing number of evenings when I kept my own company. A little triumph arrived with the first time I realised I actively wanted to be in, by myself, but I still found it impossible to stop caring for Josh. I would wonder what he was up to, wish for his well-being, hope he wasn't hurting too much and that his friends were being kind. I couldn't support him any more, but I still wanted to. Not to be doing so felt like a whole other loss of its own.

My mates were beginning to nudge me into seeing new people, get on dating apps and consider going for drinks with their nice single friends. For months, the thought of touching, let alone dating, a stranger had been utterly unthinkable. I was fully content to face the world by myself, without the support of a partner or the possible future that they could bring. But I was also starting to crave the physical company. That animalistic, almost unspeakable thrill of unpredictable excitement that comes with it, the tempting dangers that accompany those encounters. I wanted to want somebody and feel they wanted me back.

I didn't know how to get there, mind. I refused to be set up on a string of jittery dates with people idly looking to pass the time. I baulked at the act of setting up a Tinder profile; couldn't even begin to think of what photographs I would put on there, how to sum myself up in a quippy caption. I knew I'd panic at even swiping right on a stranger from the internet, worrying that they would never swipe back. I couldn't face being further rejected.

It was part of a general feeling of restlessness, the September shift finding its way into my bones. My numb summer was waning, turning into a season of surge. The need to push myself into life, and into life alone, pulsed urgently and often louder than my dulling heartbreak. After years of making London liveable and steady I wanted to force myself into its crannies, explore it like when it – and I – were new. Pace its pavements as I had

New York's. Make myself fall in love with London again, with its stratospheric offerings and grave vexations.

One Friday night, bailed on by friends and my phone rendered resolutely, infuriating quiet, I wriggled out of my pyjamas at midnight and crossed the city to a working men's club on the small chance of seeing a surprise Lady Gaga gig. While everyone else inside was drunk and gleeful and with friends, I stood alone, no drink in my hand, and watched it, felt it all unfold around me. For an hour, I was injected with that same wide-eyed freedom I'd tasted when I first moved to London, in New York when I arrived there. Keeping to familiar streets and convenient haunts had made me jaded, but I was finding possibility again. I eschewed a cab and rode several night buses home, high on the adrenaline of taking the city for my own.

•

Darkness had long settled by the time we got to the party the next day, although it was still warm enough out. Emily and I leant against the balcony railings, the rasp of whisky in our throats, and looked out at the familiar bumps of the Square Mile beyond. Gherkin, Cheesegrater, Walkie-Talkie. Further along, the persistent flash of Canary Wharf's beacon. It wasn't one of those loud, exciting things, with people squished up against one another, their flushed skin emerging from slips of fabric and the bassline of heavy, anonymous music so loud it feels primal. This was more of a drunken gathering – two handfuls of people sitting around in an overly bright flat, not quite dancing to music that was playing to inspire nostalgia. We'd only gone because it suddenly seemed like a good idea after we'd drunk a bottle of wine at the Hungarian restaurant down the road.

So when my name was called, it didn't have much to cut through. I didn't recognise him; we hadn't met before. But our industry is small, and as he introduced himself I knew who he was; I'd edited his copy years before. He was a theatre critic and had spent the day watching shows in Suffolk – proof, perhaps, that both of us were only even at that party by slim chance. I felt elevated on drink, bolshy and right-minded on the fumes of it. We chatted mindlessly at first, pushing through the flotsam of shop-talk and clumsy flirtation, but then the conversation became a thing that sparked and hissed, like green wood chucked on a fading fire. It had been years since I had performed that awkward, nudging dance, the daft and hapless game of getting to know somebody while largely pretending to be an improved version of yourself. I brought up the break-up as a kind of excuse, a verbal flag to invite forgiveness in case I seemed as broken as I felt.

At some point, Emily went home and he and I were left standing by the fridge in the kitchen. Our bodies had wound ever closer as the evening wore on. The fluorescent bar lighting bothered me, made everything feel too exposed, so I kept switching it off. I was giddy on the hit of it but hesitant, too. It took a while for me to realise that this was something that could be more than a conversation; I wanted to be near him without quite realising why, without seeing it as the beginning of a predictable string of events that would lead to only one kind of ending.

Neither of us were quite brave enough to make even the

smallest of grand gestures, and I felt somehow paralysed to try. Eventually, teetering on the brink of thin-wearing patience, we kissed, a collision of beer and anticipation that ended in shy smiles. It changed the status of things from what had been an evening into the nascent depths of a new morning. He called a cab. While we were waiting for it outside I realised I had left my phone inside. When I picked it up, it was slick and dead with someone else's drink, a digital crutch extinguished.

It was early afternoon by the time we walked out of his flat. I'd woken what had felt like a lifetime before, just minutes after dawn, not knowing what to do with the sleeping, painfully new and different body beside me. So I lay there watching the light change, watching his chest rise and fall, half-hoping, half-terrified, that he might soon wake up.

We were ejected into the fresh air of a sunny afternoon. He'd invited me to eat, but I was keen to get home, deal with my hangover and suture off the whole thing. I was glowing with the thrill of the hours before but nevertheless drenched in guilt; it still felt like an enormous betrayal of Josh to have shared myself with someone else, to have been made to feel this happy, this wanted, by somebody who wasn't him. My hair was still wet from the shower. I paced Brixton's streets feeling unusually self-conscious of my bare face, as if the sheer novelty of it all was painted where my make-up had been. The sense of what had just happened was so baffling that I hadn't had a chance to process what would unfold next, aside from a plan I'd made to call a friend that day. So when Matt paused at the corner of his street, took my face in his hands to kiss me and nervously tried to give me his number, I was taken aback – I'd honestly not even thought about seeing him again. These well-worn motions, fixed in film and narrative and life, did not feel like things that would apply to me. And anyway, my phone was broken and he'd left his inside, so we agreed to find one another online.

I walked home via the nursery. Over the years I had come to develop a small network of them, especially ones close to the flat. There was a smart one on one side of the hill in Dulwich which was good for an expensive idle potter, but I was increasingly coming to rely on the more workaday rival a mile or two away in Herne Hill. Sometimes I'd do both, but south London is irritating to cover by bus, and there are only so many plants one can carry on a bike. Croxted Road's nursery feels somehow timeless; it has been seeing in the seasons with budding plugs and veggie seedlings for decades. Inside, where the houseplants and fertiliser were sold, had the comforting, earthy smell that belongs to sheds, pet shops and grocers. I liked going in, asking questions. There was a child-like satisfaction in the brightly coloured seed packets lined up on the wall and the different varieties of bulbs lying in boxes in their crinkled skin.

It was tenderstem broccoli plug plants I brought back to the balcony that balmy afternoon, in clothes crumpled from the day before. A most optimistic purchase, as many naive ones are. Broccoli isn't particularly difficult to grow, but it does need things I couldn't provide, namely a good half a metre (at least) around each plant. Instead, I shoved my new seedlings into a plastic trough better suited to a small crop of salad leaves – the endive was still going strong – and dreamed of balcony harvests in the spring.

I ended up seeing Matt again. He was open and honest in a way I hadn't allowed myself to expect from friends' horror stories of ghosting and my fear of the dating apps. My own, relatively ancient, experiences of trying to go out with people had been litter-strewn with texts that went unanswered and mixed messages left hanging in the air.

But I liked the way his easy manner made me feel, as if I'd been covered in a fine mist of pliable varnish or had a really good blow dry. He noticed qualities I had forgotten I had, funny things that felt completely unremarkable, like my teeth, or the fact I used 'crikey' in daily conversation. He didn't so much compliment me in bits but as a whole, as if he liked all of me. After spending so long picking over myself for reasons why I hadn't been good enough, here was someone who simply, and quite literally, told me that I was good. To be able to be appreciated and seen by a stranger suggested that I was opening up a little. Capable of change, of allowing myself to tilt with the seasons, releasing some of the distrust and upset that had punctuated the summer. With Matt I felt shiny. I hadn't felt shiny without chemicals or alcohol for weeks.

I took him to the South London Botanical Institute, the kind of relic that nestles in an ordinary road in Tulse Hill. It's a charmingly archaic institution inside a Victorian house that would have housed a comfortable family had it not been filled with 2,000 books and good intentions. Allan Octavian Hume was a civil servant in India for forty-five years. When he returned to London, he turned his attentions to plants. His resolve was the very reverse of that which fulfilled the upper, exclusive echelons of early

Victorian botany: to make the study of plants accessible for all, especially the working classes. Setting up the Institute in 1910 was part of that, a place where the herbariums he kept, filled with specimens he had grown, could be used for reference by the many.

These days, the library is open for just a few hours one day a week, but for a handful of days a year the public are invited in. And we happened to be meeting on one of them. There, we padded around the two rooms that constitute its library and looked at Hume's herbaria – the place has barely changed since he died a couple of years after he founded it – before squeezing into the perfectly small and shambling glasshouse that leads onto the garden beyond. Hume had transformed the domestic plot into a botanical garden that still boasts some 500 varieties, but it had succumbed to that late-summer grogginess its neighbouring gardens had. Grand plans had overgrown; the brass plaques boldly labelling plots such as the medicinal plants and DIY borders either stood on patches of bare ground or were smothered in plants. Tree ferns stood modestly in a corner above a bank of things that had been left to take the land. A small pond gently puckered in the warm sunlight. This was a space that begged for time and reflection, and it was difficult to offer that when I was so conscious of myself, of the words I was saying, the impression I was giving. We sat on a bench, a little precious with each other, and nudged at the boundaries we were building up and taking down in some complicated, invisible routine. He made jokes about ferns that I didn't find funny, but I still let him take me for a pint after.

When I got home, the light in the bedroom was turning pink

and sank into the pillows that were drenched in it; a kind of steadying. Although I was sleeping better – no longer waking and feeling the emptiness of the other side of the bed, or feeling racked with a misshapen grief – I still wasn't sleeping much during those September weekends and they passed in hours of floating haziness. The equinox – and with it, my twenty-eighth birthday – arrived, and I sat around a table with twelve women who had helped me limp through the months before, eating ourselves to a standstill and cackling over stories involving nipple piercings. Slowly, I was learning to enjoy the loosened reins I had been granted by change, to play with them in learning how better to adapt to the new confines of my life. Not replacing what had failed to thrive, but giving something else a chance to linger in that new, empty space. What that was I didn't yet know.

A week later, I travelled up to a part of London so far north I hadn't been there in nearly a decade, inhaled a Crunchie bar and, with the honeycomb still lingering around my gums, had eight delicate maidenhair fern leaves inked into my inner bicep in a mass of tiny polka dots. I'd decided to get the tattoo six weeks earlier, sitting in London Fields with former work colleagues. One of whom, also in the midst of a break-up, had put his arm under the needle. It was a cliché, I knew, but it also felt like a conclusion to something I'd dreamed up years before, an intention planted when I'd lived in Brooklyn, and seen black marks left on the bodies of my new neighbours. I'd flirted with the idea ever since.

The image had been lingering on my phone for months but I came across it when I was collecting photos for another project:

the freshly unfurled growth of a maidenhair fern that had been accidentally severed while I was cutting back older leaves. Maidenhair ferns are precious but beautiful things that respond well to having their crispier brown leaves removed. But it is easy to be slapdash and sometimes cut the wiry stems of new fronds in the process. This one sat on the kitchen side perfectly intact and perfectly broken at the same time, like a specimen ready for preserving in an album. I wasn't sure of the significance it held, this memento mori of a few little leaves that wound up in the bin but also etched into my skin for ever, but nor did I think about it much. I emailed Martha, who was part of the loose group of art students I hung around with when I first moved to London. She had studied illustration but was rapidly gaining a reputation for her tattooing, which was as pretty as it was precise and based on the practice of Victorian etching. And when the afternoon arrived, she was swift and businesslike in her work. A job done well and quickly that would last for the rest of my days. A piece of fragile nature, cut down in its prime, had gained permanence on the body I was reacquainting myself with as much as I was everything else in my life. I adored the tattoo. Autumn was drawing in fast; it wouldn't have to be properly on show for months. I thought about who that woman might be while keeping it as a vague, unspoken mark of that summer.

friendly strangers, but had seemed like a good idea a couple of months ago. The notion of it appealed: wandering around rain-slicked streets and having *plat du jour* lunches by myself, wrapped up in a scarf and listening to playlists chosen for the purpose. I wanted to sink into those slate-grey roofs and casual curlicature for a few days. There wasn't much of that in reality – more boozy dinners paid for by some rising star's record label and a struggle to stay up to watch the headliners, who were put on in the small hours of the morning. I wound up in bed steeped in red wine, wondering a bit why I'd gone.

While the others slept off their hangovers I filled up on the plant life I could find. Discovered the small community garden that occupies a patch of space behind the Palais de Tokyo. The city was still largely green, although yellowing leaves drifted onto the avenues, catching the flurries of autumnal wind. In the community plot, beef tomatoes clung on – whether in abandonment or hope, I'll never know – with a view of the Eiffel Tower, wrapped in insouciant sky. Tried to take in the formal expanse of the Jardin des Plantes, France's equivalent of Kew, but got listless with the long, straight gardens in which they put their discoveries. The beds were dying down; a small army of gardeners were pulling out dwindling varieties and tending to the soil. The dull light that escaped between the fast-moving clouds was captured by a few remaining pink poppies, their papery petals diffusing it.

When I reached the city's flower market, Marché aux Fleurs et aux Oiseaux, at midday on a Friday, I was the only customer there – and remained so throughout my short visit. It carried an intangible muteness, this neat little place. Squat green stalls lined up in rows, filled precisely with plants and presided over by apathetic stallholders. It almost respired with pretty loneliness. I refused to be taken in with it. If anything, the whole place made me long for Columbia Road, the scruffiness of it, the thrust of it, the fact that if you don't know what you're after or what you're doing at the east London market, you will get swept along with the tide who turn up, hearing yells from the stallholders and taking in the smell of overwatered lavender.

Paris could offer certain things – champagne in every bistro open beyond midnight; elegant, chaotic parties in gentrified squats – but I struggled to find a green space with which to resonate here. Everything was somehow too trussed up to be allowed to germinate.

•

London, meanwhile, was changing as autumn blew in. Not that I saw much of it; as well as Paris, I'd flown to Berlin for a few days and had sandwiched the two trips between a variety of sofa-beds and spare rooms. My calendar became sliced up into dates and unfamiliar addresses, marked by the rattling of a red suitcase across paving slabs and up station staircases. The thing was only a couple of years old but its back wheels were already battered. By the end of the month, they would be reduced to jagged squares, leaving faint lines in the cooling dirt behind me.

The constant shuffling, and the endless thinking it induced, made me snappy and agitated. I got so fed up with the daily catch-ups with the kind people who had taken me in because I just wanted to rest instead, rather than make small talk after a new commute and a long day. So much of my brain was occupied with working out where I'd sleep, and how I'd get there, that all the social commitments I'd plugged into my diary became a chore. I'd abandon what I could, usually on the day of it, riddled with guilt and relief in equal measure. The ones that stayed in place I'd be torn between cutting short, because I simply didn't have the energy, or stringing out to put off going home to yet another new place.

I was out of practice at looking for places to rent, that was the truth of it. I was one of millions of young people living in London who was struggling to find somewhere to live, who was between places, but I'd been fortunate enough not to have to think about it for a while. A few years earlier I'd been far more practised: between the ages of twenty-one and twenty-three I'd racked up six addresses, which was fairly standard for city-dwellers my age. 'Digital nomadism' has become a glamorous way of dressing up the way that subletting and short leases and unexplained evictions have rendered us without a sense of permanence, nor even a hope of it. Most of us are well-versed in the trawl through Gumtree, the art of making ourselves sound appealing as a housemate ('tidy and chill and up for a glass of wine every now and then'). It was such an elaborate game of cat-and-mouse, and so brutally playful with the borders between truth and fiction. I'd stuck a plea for a temporary room up on a south London Facebook group, and was awash with misplaced shame when my friend said she'd seen it. The situation wasn't embarrassing, but the notion of my little pitch being examined and passed up by strangers was.

I did find a strange catharsis, though, in the sense of self-sufficiency it brought me. I imagined the schadenfreude it must have induced, that the home-owning millennial was now cadging around for a room while paying her mortgage, but I took genuine pleasure in how streamlined my life had become. I relished the generosity of old, fairly lost, friends who would take me in. I took pride in the fact I was getting better at packing lightly, of economising what I needed – the woman who sweatily dragged desperately packed luggage around London in late summer had evolved into something more sprightly and practical. I was learning to cope better with less – fewer clothes, less stuff – in order to allow more jaunts, more conversations, more love into my life. I was a living example of the adage our generation thrives by: that experience trumps object.

Quite without warning, I was tumbling into Matt, too. Into his inbox, into his arms, into his bed. We defied the socially acceptable ways of dating among people our age; to see one another every now and then, perhaps on a weekly basis, and possibly while keeping a few other people on the go until explicit monogamy had been stated. Instead, we collided as if we'd shared the same Tube carriage at rush hour and decided to spend the rest of the day being tourists in our own city.

We'd drop each other text messages in the afternoon or evening, be having a drink somewhere small and dark and late a few hours later. I'd be out doing something else and find my plans changing, Google Maps instructions diverting, as I made my way over to his flat in the depths of the evening. I wouldn't always let on to my friends; outwardly, it seemed so intense. They'd raise an eyebrow, say, 'You're seeing him AGAIN?' It felt almost appropriate to be a little sneaky about it.

At first I thought this spontaneity was some kind of well-orchestrated performance, but it was actually just how he functioned: a ball of chaos that bumped through life fitting things into days that sprawled and squeezed. While I planned and fretted and analysed to the extent that I would examine the spaces between our text messages, to see if he was deliberately holding back from replying like I was, he just seemed to be. He would come out with daft and lovely observations about me as casually as he would ask if I'd like a cup of tea. I was amazed at how we jigsawed together, how much space he started to take up in my mind. I got hooked on him.

I didn't let him know, though. I kept myself at a distance, emotionally at least. Fell into that trap of playing the cool girl, who seems unfazed and breezy and light, when inside I was all excitement and confusion and constant, futile questioning. He didn't, for instance, know any of the details about my ever-switching postcodes, because I didn't want to tar the clean alchemy of what lay between us with the messy tedium of that. There was also the fact he compartmentalised things: if he was working a lot, he wouldn't get back to me, but when we were together that was all there was. 'Where did you come from?' he'd ask, not needing an answer as he let his fingers drift through my hair. I wouldn't give him one, either, jumping up to throw on yesterday's clothes and head out to work late because it was always more tempting to stay next to him, just for a few more minutes.

To be with Matt was to enter a bubble that glazed off the weekly upheaval of home and the tumult of my mind. I was still playing someone else, but she was someone I felt improved me, a person who didn't feel as hollow as the one who politely invaded the spare rooms of people I'd last properly hung out with several years before.

I was always touched by how the break-up had erased the distance between me and some old friends. Holly put me up, and we'd barely had contact since the age of twenty-three. When she arrived in New York, I showed her the little I'd learned in my first few weeks there. We took each other out, two British girls exploring the invisible boundaries of our freedoms. Once back in London, our lives shifted away from one another. Now, five years later, Holly lent me her space – no questions asked – honouring a silent contract we'd made on the edges of our girlhood to look out for one another in uncertain times. Catching up became an unspoken game of spot the difference, in which we both reverted to the people we had been when we first met and worked out how we had changed. Instead of the boys and bars of Brooklyn, Holly and I were preoccupied with a new maturity: she was doing up her kitchen and learning Japanese; I was sleeping on a deflating air bed with as much dignity as I could muster.

Four months had passed since Josh and I broke up; people's lives and interests had moved on from comforting me, and that was fair. There was an unspoken suggestion that I should be coping now, and I felt it too. But I still couldn't ignore the fact that bits of us, of him, were drifting from me: our own silly lexicon, some of the in-jokes. As the crisp leaves turned to wet mush in the gutter, I felt that gentle decay of what we had made together. I couldn't work out if I still loved him or not, it was a concept too gauzy to grasp, but care, oh, that was there. I cared for him just as much, only it wasn't fair or right for me to show him. Instead, I swallowed it down, made peace with those longings that interrupted my day.

•

Like those people who vacuum their way out of the door before going on holiday, I put in as much time I could muster on the balcony before I left it. Being able to breathe it in left me feeling just a bit lighter. I'd become dependent on those minutes outside, reading the hidden language of what was growing there, what the trees were doing beyond it, the movement of the skies and the clouds within them. It was a vital counterbalance.

The last morning before I left had been an industrious one. I was up minutes after sunrise, a measly handful of hours after I'd gone to bed, with a rabid hunger to shift the balcony around and sleep dust and make-up still lingering around my eyes. It was an ice cream dawn, pink streaking a sky that dwarfed the city and glossed the millions of glass panels that comprise it. Soon I would be down among those roads, back on the ground with the sirens and the SpareRoom ads, waiting for buses and shuffling against other people caught in boxes where work and relaxation happen in an unfair balance. To have it all was so often to be removed from the physical reality of things: to live a life that happened mostly through screens.

But here, in this little concrete box in the sky, lay defiant life. Even though I was being wrenched from it, even though I didn't know how long I would be looking after it, the thought of abandoning the balcony – letting it go to seed – was impossible to entertain. To nurture it, to work with it, was to physically engage with it. As I shuffled the heavy containers around, puffer jacket over pyjamas, I could feel my muscles working, my hands smudge with the cold dirt in the corners left behind. I felt my heart rate creep up, saw my breath paint itself faintly against the warming air. It was an exercise of purpose, a way of doing things that I didn't, for once, have to plan or analyse for anything other than itself. To garden that space was as instinctive as dancing; one move followed the other through feeling and necessity alone. I rarely planned the minutiae of it, of how much to water or what to pull out, I just did it when it arose.

The best guidance of where to start, though, is with the seasons ahead. It is easy to let summer slide into the end of September without doing much gardening, but the prospect of returning in November without having prepared for the winter ahead was an unpleasant one, even though at that point the future was so muddy that I couldn't imagine what spring would look like or where I'd be witnessing it.

The *Pelargoniums* that had been filling the window boxes with a foam of petals had, a few days before, been uprooted onto newspaper on the fold-out melamine table I used as a potting bench. London is warm enough to get away with not doing much to *Pelargoniums* over winter. Plenty of them will last through the cold in window boxes and put on a good show the summer after, and in years previous I'd just moved the pots to the wall, where it was warm and sheltered from the extremities suffered by the ledge. A more common argument is just not to bother – *Pelargoniums*, commonly sold as geraniums, are so cheap as to be easily replaced. But without a ready compost bin it felt a bit sacrilegious; why vanquish such determination to survive?

I did fancy a change, though. The plants had been a Columbia Road Flower Market bargain some months before and I was still practising the different arts of doing things properly. In this case, lifting and cutting them back before repotting with fresh soil, better to help them store the energy needed to return with a vengeance once the gloom had passed. *Pelargoniums* will make rangy branches in the pursuit of light, and the warm summer had left these straggling beneath their flowers. I forced my fingers into the overgrown window boxes and wrestled out their expanded root balls, making the meagre quarter turn needed to place them on the paper behind me. The smell of cooling earth mingled with the invigorating pepper of the leaves. Secateurs were taken to the stems, leaving behind a fist-sized head of thick stubs, from which new leaves would grow. They looked both vulnerable and sturdy at the same time, as if they'd gone out in a thick jumper but forgotten their coat. I pushed them down snugly into newly washed pots and lined them up on the table against the warm brick wall of the flat.

A flurry of fat white cyclamen landed in their place. Their elegance defies their hardiness. Those inside-out petals may look like bone china but they'll weather a storm, only succumbing to rot if the soil doesn't have enough drainage. When the flowers fade, in the New Year, the leaves – like rounded hearts, smattered in a spiderweb of pale green – are allowed their moment in the spotlight. They went in with other plants that don't mind having wet feet and would be able to look after themselves in my absence: ivy, which will fill a space with voracious green if the soil is damp enough, and violas. The three of them are cheap, none fancy, and all capable of bringing light in a grey winter morning.

I'd been given a large bamboo, which was a fun treat. After the ferns that swelled in the darker corners of the balcony, it carried the beginnings of my development into a braver, more ambitious small-space gardener. Along with a few other shrubs, it would become an essential tool to transform a concrete space into a green one. But one does not begin to garden with things taller than oneself; it's like letting a large, wet dog run into an antique shop littered with precious trip hazards – you need to be sure you know what to do if it goes wrong. That morning I made the first steps towards putting the dog on the leash, and dragged that hulk of rustling leaf and stone pot two metres across the balcony.

The sedum, which had turned chocolatey – and proved still as sweet to the bees – had to accommodate this new neighbour. I admired how their contrasting leaves sat together, slowly accepting that they could sit side by side, relishing the potential rule-breaking of letting two plants collide beyond the remits of a window box. I worked quickly and against the chill of the October dawn, fingers not minding the cold and the muck and the strain. The productivity of it, the speed of hatching a plan and seeing it through seconds later, cut through my sleep deprivation, my hungover gut. It silenced the droning to-do lists that built up in my brain.

Learning how to look is vital in gardening, especially for someone as impatient and haste-driven as me. But the balcony supplied near-endless visual satisfaction; I'd hunt out new growth, watch the light play on the leaves, catch the shadows that fell on the concrete floor. As my father did standing by our kitchen window, I'd observe and make endless mental lists of things that needed to change or growth I'd hope to see while my focus fell soft and heavy. It was near-impossible to do just the one quick job on the balcony. A bit of deadheading leads to watering, to tidying and pottering around until the five minutes initially allocated telescopes into hours and my hands – usually brushed roughly on whatever covers my lower half – are stiff with cold and my mind soothed.

Katharine S. White – née Angell – was the first fiction editor of the *New Yorker*, the woman who discovered and championed Nabokov and Updike. But she applied her gimlet eye to seed catalogues, too, and at the age of sixty-six wrote her first gardening column, 'Onward and Upward in the Garden', in which she treated the plantsmen and nurserymen who collated their wares in pamphlets as if they were the next literary sensations. Even though White (she married one of her protégés, E.B. White) changed the shape of contemporary fiction with her day job, her 'favourite reading matter' was seed catalogues.

This is a charming thing, but my favourite Katharine S. White fact is that, as her husband wrote when he published her columns in a book after her death in 1977:

> [She] had no garden clothes and never dressed for gardening. When she paid a call on her perennial borders or her cutting bed or her rose garden, she was not dressed for the part – she was simply a spur-of-the-moment escapee from the house and, in the early years, from the job of editing manuscripts . . . I

I began to shift my existence outwards as the places where I slept became increasingly temporary. My attention kept moving outside. Work, the job I'd been driven towards for a decade, started to lose its hold on me. I just couldn't care about the Mercury Prize winners or the latest surprise album drop any more, I struggled to keep new music in my brain or nurture a desire for it. Somehow it couldn't express meaning in the way I felt things now. I sought something more resilient, something that could last and grow. Every time I clicked through one of the dozens of daily press releases screaming about a fashionable new band I'd feel the futility of it all even more sharply; pop star album campaigns would be over in weeks, there was no room for anything to grow or develop.

I became increasingly content to go through the motions, would get things done as efficiently as possible while clock-watching for hometime. The minute those hours were over, I'd leave in a flurry of guilt and rebellion and freedom at clocking off when it was permissible in a culture where overtime was the norm. But I was hungry to be out, desperate with it, needed to leave my desk and discover the untapped green of the city. What had started as a curiosity – to see what was growing and how, to hunt for green or turning ochre against the grey, to look for clues of what would happen next in these plants' secretive, silent lives – had become a kind of compulsion. I found myself craving it, feeling confined in in spaces where there were no plants. I'd silently suss out where urban nature lay, like a smoker cadging for a cigarette. I'd mentally note the front gardens I would pass, whether there was a way to walk home through a park. It became a kind of treasure hunt, with bonus points for summer flowers still stoically blooming even with the arrival of Daylight Saving, the hydrangea blooms elegantly fading against the shortening days. All of life's processes were on show. Autumn was well underway, the leaves were burning and falling, leaving the strong skeletons of trees to stake out the colder months, to bide time through another winter, one of dozens they had witnessed.

There was constancy in these snatched sights, far more constancy than in my own life. And I relished it, its solidity, its independence from the life we humans built around it. I was removed from home and wrenched from the predictable path I had thought I was on. Perhaps in plants I could find such a steadfast way of being, far beyond a life I had come to expect for myself.

•

Parks are a curious phenomenon in the countryside, something that you drive to, a site of play equipment but little else. And they're not really parks, but 'recs'. Our village had one that was overgrown and full of warning, that we weren't allowed to play in. And so they were always something of a novelty in my childhood. A trip to my grandparents' in Reading meant a post-lunch walk to the rec – arguably the highlight of the day, sometimes beaten by the mounds of heavily buttered pasta my gran would make. But I was slow to pick up on the notion of them. Like northern slang and having two different types of potato with a roast dinner, parks – the joy and purpose thereof – were another welcome interruption to my sheltered adolescence that arrived when I moved up to Newcastle.

Having never lived in a town, much less a city, the notion of having to be provided with an outdoor space for the pure pleasure of it wasn't a concept I understood because I had never had to think about it much. The naivety of this struck me during Freshers' Week. I was nineteen by a day when I moved into my new home, in a now-demolished block of halls opposite the Town Moor and, further down Richardson Road, Leazes Park. Many were bemused by the existence of the Moor, a vast welt of grassland that, at 1,000 acres, dwarfs Hampstead Heath and Hyde Park put together, but I found it a comfort. The grazing cows, which students would attempt to 'tip' – or push over – on the way home from nights out, reminded me of home. Stories would proliferate about flashers and bad men; my flatmates would run around it in the dark evenings in a pack to ward off danger. But for me it was an invigorating place, so large that to stand in the middle of it was to be removed from the city; the grass heads would whip against each other, birds would dip and dive for the insects below. When the wind blew, as James Taylor would say, it could turn your head around.

Leazes, though, was a novelty. I was charmed by the black iron railings and the grand stone gate posts; I found the bowling green bemusing and the boating lake fancy, although was less enamoured with it than my flatmates, who would race one another across its murky waters in the depths of the night.

I wasn't in the city to see it swamped during the summer holidays, but instead occupied the park in its quiet hours. Leazes Park was always open, which was why, along with the Moor, it attracted a reputation for untoward activity. My favourite time to walk through it was in the minutes just before daybreak, when the skies were at their deepest and heavy with birdsong, a sound so sweet to ears ringing from club music. One Sunday night we bust out and lit sparklers there, spinning on roundabouts through the clouds of sweet smoke. In those first surprisingly warm afternoons of spring, while everybody else was at work, I'd sit under the budding trees. At its most perfunctory, Leazes was a thoroughfare between home and the pub, maybe part of a round route back from the market or some of the less-frequented lecture theatres. Even then I relished it; the opportunity to look at the ducks, witness the movement of the seasons and idly chart them against the rate of my independence and what I'd learned that week.

Leazes was the product of good intentions and a difficult birth. It came from a petition to Newcastle Council in 1857 by 3,000 working men who wanted 'ready access to some open ground for the purpose of health and recreation'. Nearly twenty years later the committees and campaigns paid off. A park was opened on part of the Town Moor. The ribbon was cut on 23 December, an early Christmas present for everyone in the city. Cities may take over the land but it's only a matter of time until people start to put the green space back into them.

It ended up being the first of dozens of parks I came to fall for over the course of my twenties. In Newcastle we were more spoiled than we realised, what with the Moor and Jesmond Dene. New York boasted the High Line and Central Park among Manhattan's quainter offerings, even if Brooklyn was less well-served. But London beat them both, because green space was built into the very skeleton of the city as we know it. They remain part of its being now – nearly half of the city is green, while the average in cities is around 24 per cent. It's even become the first National Park City, a notion that is largely irrelevant to those who don't understand it and feels like a radical achievement to those that do.

Parks are referred to as the 'lungs' of the city the world over, but the term emerged in London in the eighteenth century along with England's public parks. The city was a cesspit – in many neighbourhoods, quite literally. The masses who had moved to London to work put up with squalor in the wake of the industrial revolution, living in cramped, filthy situations. The many inadequacies of the city's infrastructure are well-documented and grisly – meagre sanitation and overcrowding led to disease. In order to better understand how to improve matters, London's medical men thought of the city like a body: the water of the Thames and the sanitation system it passed through was compared to the mass of arteries, vessels and capillaries that make up a circulation system. Faced with the disease, specifically cholera, that people (wrongly) believed was carried in the air, the same logic proposed that London needed lungs, or parks: areas of green space and clean breathing that would bring oxygen into the body of the city and remove the filth that lingered in its air. These spaces would afford the confined working classes the opportunity to enjoy the health benefits of a good walk in an area of nature – which the middle and upper classes had been luxuriating in for decades.

London already had potential as a green city: our monarchs had long been ringfencing whole swathes of it for deer hunting and other aristocratic pursuits. These became known as the Royal Parks, which were given over to the public by a long-awaited act of royal generosity in 1851 and still exist today. They were enjoyed then as they are now: public spaces for promenading and horse riding (although these days we prefer bikes). But some fifty years before, at the turn of the nineteenth century, the parks were under threat. There was money to be made from building houses and the vast expanse of green space offered by Green Park, St James's Park and Hyde Park was tantalising to developers. In 1808, a motion to build eight 'of the most expensive houses . . . in the metropolis' in Hyde Park was voted against in the House of Commons, after a speech given by William Windham in which he quoted William Pitt by calling the parks the 'lungs of London'.

To someone who has watched London become a viciously fought playground of housing developers, Windham's arguments feel eerily prescient. According to the records in Cobbett's Parliamentary Debates Vol. 11, Windham said that it wouldn't just be eight houses:

> If the buildings were once begun, he was convinced the system would go on. These eight houses would not be the last . . . the power of the vegetation would be completely destroyed. The park would no longer be that scene of health and recreation it formerly was.

Windham gives an example, a man walking from Whitechapel 'on a Sunday evening to get a little fresh air' only to be greeted with 'nothing but houses. He would most probably think he had seen enough of these in the course of his walk.' To build the houses, Windham argued, would result in 'the destruction of these lungs'.

Twenty-one years later the concept cropped up again, when John Claudius Loudon, a man with just one arm and dozens of ideas on how green space would benefit humankind, published an essay on the benefits of parks as 'breathing spaces' for the masses. He was the publisher and editor of the *Gardener's Magazine*, which would carry his dogged campaign for more public parks in British cities. The notion of London's parks as lungs was used over and over again by officials who consistently pushed to keep the city green – both by preserving the parks London already had and by ensuring more open spaces would be built as the city expanded. As an article in the *Spectator* noticed in the mid 1840s, lungs are among the first organs to develop across the animal kingdom. And so it was with London: the parks – the lungs – paved the way for more streets of houses, more shops, town halls and more gardens, too.

Windham's speech took place at the start of a century that would see dozens of parks opened across the city and the country more widely. The first was in 1840, when a textiles magnate named Joseph Strutt commissioned Loudon (also a horticulturalist, garden designer and expert in trees) to create Derby Arboretum to cut through the smog of the industrial midlands town and give its workers somewhere to experience nature. Two years later, Victoria Park opened in a sprawl across Hackney, east London – so that man from Whitechapel no longer needed to schlep across the city to take fresh air. By 1852, twelve parks had been laid out in London, partly inspired by the horrifying revelation by a parliamentary committee that there was 'no single spot reserved as a park or a public walk' along the five-mile stretch between Vauxhall and Rotherhithe. Southwark and Finsbury Parks were among them. The 'darkest and dreariest' fields on the river-banks opposite Chelsea – a site of dog-fighting and duelling in the first half of the nineteenth century – were transformed into Battersea Park. Like Leazes, they shared similar trappings: a band-stand, a lake, trees and spaces for people to relax, take a breath, go for a stroll with the family rather than sit in the pub.

While they were arguably served better by their country estates, London's upper classes nurtured the same desire for green space and clean air, too. Garden squares began to pop up in the seventeenth century but rocketed in the eighteenth and nineteenth. These dreaming pockets of the city placed terraces of tall Georgian houses around squares of gardens. The first essay Loudon published, in 1803, was about how these squares were planted; he suggested plane trees rather than ferns and conifers, and two centuries on millions of people still suffer from hayfever and admire the dancing light through their leaves as a result. It was in these refined new spaces, the Whig aristocrats who were largely responsible for bringing them into existence believed, that people could converse and communicate, meet one another and exchange ideas.

These intentions may have been made by grand, privileged men with little experience of the lives which they argued for or sought to improve but they cheer me nonetheless: a powerful state reclaiming and protecting land for the use, health and benefit of everyone. The plans didn't always work out: the bacteria that caused disease didn't travel on the air, after all. The lungs became clogged, especially in cities where there was too much pollution for a few small parks to cope with, and diseased. Acid rain in

Manchester caused the plants to die; in Regent's Park, the fleece of the sheep that grazed there blackened. The garden squares became ever more elite, fenced off to all but the keyholders that lived around them, seen by others only in peering through railings and scenes in films.

Crucially, though, the lungs remained. Public campaigns reached a groundswell and in 1866, an Act of Parliament was passed to protect commonlands – such as Hampstead Heath – for public access. I am so grateful for that. Parks have been the percussion to the cacophony of my twenties, a decade lived in cities. They have given my life a steady beat. They may have been founded as a retreat from far more physically gruelling work than that I have put my body to, but the release remains. We seek parks for openness, for space and leisure far beyond the luxury of a patch of grass large enough to splay our bodies on during the sunshine-laden days. Parks offer room to clear the head and de-fuzz the mind. A patch of land dedicated purely to being away from the confines of our world, whether those be in a mine or a factory or in front of a computer screen. There is a nurturing rhythm to those iron railings, a freedom to the creak of a gate open in all weathers and seasons, to all people and all states of being.

The first London flat I lived in looked over Peckham Rye Common (farmland bought by London City Council for the public in 1894 – previously, according to William Blake, the location of trees filled with angels). When I first went there it was covered in snow and seemed ghostly in the December dark, an eerie nothingness after the hubbub of the market by the station. I was drawn back into it as winter turned to spring. I'd jog –

badly, slowly – through its paths and watch the normality of life unfold on its football pitches in a city that felt so alien. Like Leazes, it was a typical Victorian park, and I poured my loneliness and cash-strapped boredom into it – sometimes just by looking out of the window. Years later, I can trace my younger self in it, knowing the paths that I trod, hearing the conversations held there. A couple of weeks after we broke up (in Burgess Park, another, less-lovely expanse further towards Elephant and Castle), Josh and I met on the Common. We sat under the trees, enjoyed the comfort in seeing one another again as much as the oddity of it being on such different terms. There have been haphazard picnics and heatwave hangouts, leaf-crunching walks and early morning strolls; the best part of a decade has been spent drifting back to that space.

There have been so many more, besides. The late winter evening we had to clamber over the six-foot railings because we got locked into Dulwich Park. The confounding expanse of Victoria Park held revelry and hungover Sunday afternoon party post-mortems when I lived in Hackney. Watching the sun move through Nigel Dunnett's Beech Gardens at the Barbican, where the perennial planting offers something new all through the year. In the nascent days of our romance, I'd pop out of my Mayfair office and meet Josh for lunch in Green Park. Postman's Park, that little-known haven in the Square Mile which memorialises on Victorian tile those who have died performing heroic acts, has frequently offered respite from that unlovely corporate schmozzle. Berkeley Square would play host to long, hungry conversations about career ambitions with a fellow intern who became a good friend and a dazzling journalist. Hyde Park is preserved in my memory as a weekday refuge on a hot afternoon aged twenty-four and feeling like I'd bunked off school. Soho Square has always been a grubby and faintly guilty choice but all the more fun for it. It feels wrong to be there without a can of something; it feels wrong to be there alone. I've often threatened to scale the spiked bars of the garden squares in Belgravia, felt myself pulled back by my jumper at the hands of more sensible company.

Because if a park isn't open, then it asks to be quietly conquered, in the falling dusk, or the darkness before the dawn. When I worked in Bloomsbury, Fitzroy Square and Tavistock Gardens would jostle for pole position as a lunch-break venue, depending on whether the kind, rebellious keyholder at the former fancied 'absent-mindedly' leaving the gate off the latch. Often they did, and I would relish the small miracle of being let in on another life, another's private garden, so different from those I grew up in, just for a few minutes. Ruskin Park, which I only discovered when I moved next to it, became a kind of mirror to the life I made with Josh. It sheltered our happiness and weathered our storms. As our relationship became a brittle, fragile thing, held together by little more than denial and determination, I often went to Ruskin Park alone. Spent quiet weekend mornings watching the ducklings swim and finding quietude in the Victorian walled garden, where the herbaceous borders would stretch with spring fervour as I tried to find space and reason for my hot, angry tears. In the weeks and months after we ended, Ruskin Park became an artery. I walked and cycled through it more than ever before, one iron railing to the other, shrugging off one layer of my womanhood as another formed.

I sometimes feel that the most significant of the small adventures of my life have been taken on park benches: first dates, fall-outs, long-awaited reunions and the giddy, perambulating hours that comprise a friendship. The space of a park is undeniably public but brings with it an innate neutrality; the ability to say things that would not be fair or permissible at home or in a pub or cafe. For their creators and protectors, lungs literally gave us more oxygen. But they arguably did that in all sorts of ways: they give us the ability to clear the air and breathe life into new possibilities. They give us the space to better understand our situations, to grow into them. To look around a park is to see all manner of life compartmentalised. Scenes witnessed during a swift stroll through the narrow, oft-overlooked strip of a park that is Compton Terrace in Islington's Upper Street: a bedraggled group clutching cans and conversing in the impenetrable way of people who know each other very well and who are slightly pissed. A pair of women furrow brows over a dispute that is either underway between them, or being rehashed for the other's unpicking. A man sits down, wearily supercilious, while a woman leans over him, trying to reason him into feeling. I wondered if they had noticed the couple mere feet away, kissing unabashedly, heady in the haze of blossoming romance. A Richard Curtis film played backwards.

In parks, traffic stills and time can telescope. They are both intrinsic to the city but removed from it. I feel I've done my growing up in them: the nervous meetings and awful endings and delirious, wasted afternoons, doing little other than watching pigeons dick about. My city addresses changed – upgrading one grim bathroom for another, and then another with less mould and more expensive shampoo – and with them my interests; favourite pairs of jeans; tatty paperbacks in different rucksacks; new tyres on old, newly handed-down bikes, as I rattled through my twenties. Countless miles of iron railings have witnessed the warm-ups and aftermaths of job interviews, the dashes of heart-break, the raw upset of crumpled friendships. Life has shifted and changed, and even when it stilled, I still parked. Even in those in-betweeny years when things were calm and quiet and I strove so hard to make it work only to come out the other side with nothing but dust and pain to show for it, the parks were there – even if I wasn't in them. I'd look over at the trees of the nearest one, from the balcony, hear the strains of the jazz band strike up as I sat and read, feet propped on the edge. It wasn't really a space we explored much, kept in our nearby bubble. There were some summer days when we took a blanket out but we would wind up inside soon enough. Now I was thrust back into them again. Life had shifted and changed, but the parks had maintained a steady, necessary rhythm: the predictability of inhalation and exhalation. It felt like a homecoming.

Then there are the in-between spaces, the ones neither legislated nor owned, the ones that exist between blurred boundaries, in the middle of the privacy of a tended garden and the grand public stage of a space for all. The spaces that are seen more as opportunities, as places for things to grow. October was full of these; I sought them out too.

In the middle of the month I went to Berlin for a week or so. The German capital had proved a site of escapist pilgrimage for the best part of a decade; friends older and braver than I had moved there during and immediately after our university years. Our generation was the latest of many who had looked to Berlin for hedonism and freedom, for life beyond the restraints that penned in life elsewhere, with unruly hours and wild clubs and relaxed attitudes. I never moved there, although I thought about it a bit. But enough of my friends did, finding rangy apartments large enough to accommodate an extra sleeping body on the floor, and so I went out most years, and always with Heather, my best friend from university. It was a habit that, with hindsight, allowed us to chart our twenties by the trips. We graduated from sharing a mattress on the floor between three and between the hours of six a.m. until midday to being granted the luxury of a double bed in a spare room. Our hosts grew up, from taking dossy Erasmus years instead of sitting exams to settling down and building lives. With every trip, we'd cram into the old-fashioned Photoautomat picture booths that appear on the streets of Kreuzberg. I still have the black-and-white strips. The flash is too bright to see if we've aged or not, but the passing years can be measured by changes to our hair and make-up, the seasons sensed in fur coats and sunglasses.

The clubs and the amount of time we spent in them would alter, swap for playing games of Scrabble and eating as many meals as we could fit into a day. What we would always end up doing, though, was walking. It allowed us the kind of side-by-side space to have those conversations free of boundaries.

Berlin may have a reputation for ugliness, with its big, semi-empty concrete buildings and a river riddled with the scars of division. But I always think of the trees that line and soften the city's broad avenues. They're mostly linden trees (in England, we know them as limes), but maples, oaks and plane trees also furnish Berlin's streets, turn them into tunnels of foliage. In mid October, the colours had only just started to shift. It was cold enough for coats and woolly hats, but the leaves were caught between green and red, not yet beautiful in death. Katie lived near Puschkinallee, a street that spans Treptower Park and was planted with 1,200 plane trees, tightly packed together in four rows, over the course of three years in the late 1870s. On bright days, the foliage would keep the light captive, disseminating it on the graffitied walls and tidy pavements below. But Berlin fits pewter skies and damp dawns like a well-worn, thick-soled boot and on these days, it is easier to see how a more grounded, a more man-made nature has been planted into the gaps in the city's concrete.

It's a city where the pursuit of pleasure is rife but rule-breaking is frowned upon. Nobody, for instance, crosses the road on a red symbol (the stylised walking man – *Ampelmännchen* in German – in the former East German side of Berlin, a far less fun symbol on the West) and cycle lanes are strayed into only by tourists. All of which makes the appearance of the seemingly impromptu community gardens in Berlin's neighbourhoods more intriguing. The pocket squares of ground, fenced off by cobblestones, that sit beneath the trees play home to nasturtiums and cosmos. Others boast edibles such as pumpkins. Katie said that there was no paperwork involved; people just plant stuff when they see bare ground.

There are more formal community gardens, although even these have a more bacchanalian air than their English cousins.

Prinzessinnengärten, next to Moritzplatz Station, is a huge communal farm that was founded by amateur gardeners in 2009. Katie showed us it as an afterthought – she had wanted to buy some stationery at a shop nearby – but we lost the best part of an hour there. It had rained that morning and everything had a slight air of recovery. Puddles mellowed underfoot, the smell of wet soil stirred as rain-dewed leaves brushed our thighs. I marvelled at the ingenuity of it all; although there was verdant, edible growth everywhere (a brief, possibly mistranslated, list included chard, purslane, basil, kale, radishes, bok choy and mustard), the whole lot was grown in containers: milk cartons, rice bags, the primary-coloured crates more frequently seen carrying Berlin's glass bottles to be recycled. It meant the crops were transportable. In theory, any space lacking in greenery could potentially gain a small flock of growing cabbages in a horticultural emergency. This was growing things for the primary purpose of sharing and enjoying them, plants nurtured because there was an abandoned hunk of empty concrete and it made more sense to have green things there than grey – even if there wasn't any soil to grow in.

A forty-minute walk to the south took us to Tempelhof, a massive expanse of old airfield where kites fly and people are shrunk by the sheer enormity of it all. On a bright day, the strange nothingness of it all – 865 acres of just grass and concrete, 50 per cent larger than Monaco – can make it otherworldly. It might be Tempelhof's haunted past: it was the site of the only official SS concentration camp in the German capital. In the late Forties, that same damaged land became one of opportunity as children would come and collect the sweets thrown out of planes by

middle of urban sprawl. Berlin fought for its expanses of windy green, made political playgrounds on the pavement and smothered its streets in towering trees. Paris wrapped its parks up in bows, encouraged them to be chic. But London made itself open, made walking and thinking and playing easy. If you were patient and determined, you were rarely more than a bus ride from a bit of woodland in London. Outsiders knew London for its hustle and its expense, its eccentricities and Tube. But those who knew London loved it in spite of its filthy streets and chasms of inequality, would have a favourite park somewhere, a part of their memories steeped in green. I'd done enough European exploring; I wanted to be back in town, to trawl its parks, applaud the way the streets look when they're smothered in falling leaves.

I took the Tube down to Matt's as soon as I was back. It was a Sunday lunchtime, and I was weary, days of wandering and nights of parties left little time to rest. But he dragged me out to his favourite local green space, Brockwell Park, for a walk. It was close to me, too, but I didn't frequent it – in London, we are so spoiled with parks that it is easy to ignore even the good ones nearby. As we crested the hill that occupies the park's centre, we discovered a community garden. Not a collection of beer crates like Prinzessinnengärten, nor a patch of land gleaned next to the pavement like the Palais de Tokyo spot, but a proper Victorian walled garden, with glasshouses and signs and plants and life. It felt like a secret that had been hidden from me for too long. As we walked around it, I pointed out the different things growing: the purple kale; the brassicas; the herb patch, rubbing leaves between my fingers and lifting them for him to take in the scent of rosemary and marjoram. I knew I wanted to come back, that I had found a new space, a new green lung with which to breathe.

•

The baby was hardly a week old when I went to meet him, one evening that had grown dark with the shifting of the clocks. Hannah's house was warm, glowing soft and yellow. It smelled different somehow – the artificial comfort of laundry from the drier mingling with something more animalistic, something new and bloody. She sat on the sofa and handed him over shortly after I'd joined her, taken off my coat. He was too solid to squirm, tufty head resting on my bicep, little mouth looking for something to suck. I nudged the crook of my knuckle in there, felt him pull strongly on it. The whole thing was dizzying, a full-bodied rush of heat and love; pure happiness for them and what they had made. We all marvelled in it, in him. For half an hour, a month's worth of running around stopped. With it, all the different kinds of things I was trying to be shrank away. Instead, I found an honest, undeniable belonging. One that didn't need dressing up.

NOVEMBER

I'VE OFTEN THOUGHT OF GARDENING as being like a language. It cultivates terminologies: rootstock, cutting, grafting, perennial, hardening off, leggy. Words that are understood by those who use and need them but not by the uninitiated. Then there's the Latin, and the codes it makes that means those who can break it will be able to work out the whole family lineage of a plant in one short sentence. A cowslip, for instance, may have been called a cowslip for centuries – ever since those pretty little flowers cut a pale yellow dash against the bovine manure and bogs they grew among. But scientifically, it is a *Primula veris* (*veris* meaning 'spring'), connected to the same family as primroses (*Primula vulgaris* (*vulgaris* meaning 'widespread'), often found as cheap bedding plants in early spring) and auriculas, delicate but hardy mountain plants whose mysterious truculence and beauty has kept generations of gardeners curious.

For me, gardening's appeal lies in the fact it has to be unravelled. A language gives the speaker many different ways to communicate things, provided they stick roughly by the rules.

But plants work that way too. An overgrown and bedraggled pampas grass outside the suburban home will, certainly to my parents' generation, be strongly associated with the practice of swinging. Fast-forward a couple of decades and pampas grass has become fashionable (and firmly removed from polyamory, which no longer lurks in such secrecy), its feathery seed heads lending weight and movement to elegant and dramatic floral arrangements, becoming an essential accessory among interiors bloggers. Over the years trends and society will change a plant's meaning; they fall from fashion and bubble back into vogue just as our clothes and sofas do. My mother can't stand dahlias, partly because she was put off them by my grandfather's efforts. She says his always emerged in 'that violent yellow or mauve colour', a perhaps underwhelming reward for the 'huge to-do' over the lifting and storing of the tubers at the right time in late November. There would be 'great sadness' when the inevitable happened and the tubers were found rotten after months spent in the shed. The women of my generation don't share her dahlia distaste: the soft taupe petals of the Café au Lait variety have filled millennial brides' bouquets for the past decade. The structures behind their wild resurgence? Pinterest – and a hit of Sixties nostalgia.

As words have complex etymological histories, so plants have their rich backstories. Explorer ships and conquest; money and persuasion; smuggling and an absence of scruples; pride and keeping up appearances – these are the mechanisms that placed plants in our possession. Centuries of fascination and passing fashion saw the moth orchid (*Phalaenopsis*) removed from the tropics and delivered to these shores as a luxury, only to reach such a level of mass production they now sit under strip lighting on supermarket shelves. And those are just the horticultural languages we have managed to decode, the stories that have been told once the humans got involved. Botanists are still trying to translate the histories of what happened to the plants before we found them, and the ones of those yet to be discovered.

Family histories can be as murky and as complex. Details get lost in time, bad record-keeping and social nicety. The record states that in 1839 a Frenchman named Jean-Baptiste Vincent returned to London as a courier. But we don't know if he actually married Isabella, a cook who, according to the records, was widowed by the time she was thirty-six and mother to my grandfather's great-grandfather. Nor do we know if Vincent fathered her son, but the paltry paperwork that remains says that he did, and so this man is believed to be the reason for my surname. This is a story too distant to take a proper grasp of, partly because it was a long time ago but also because I don't know who these lovers – my ancestors – were as people, beyond the curlicues of their handwriting in a chart. I think of Isabella sometimes, especially when I'm walking past St James's Church on Piccadilly, where her sons were baptised (it's thought only one of them was Vincent's son, and therefore my ancestor), or through Mayfair, where she worked in those grand houses. But her life is too far removed from mine to draw lines to beyond sheer coincidence or the notion of walking where she did a dozen-and-a-half decades before.

Instead, we reach for the less tangible things to make sense of who made us. The kind of stuff that can't be found in an address or a census record. The rise of a cheekbone or the way someone moves their head when caught in thought. A pattern of speech or how someone always flicks off the switches on plug sockets, even if they are empty – just as their grandmother did before them. Families pass down more than genes; there are memories and habits formed half in blood and half in home. Things captured on clumsily held VHS cameras as much as they are in temperament and tissue.

People have been cross-breeding plant families, or varieties, for centuries. A playing-god acceleration of what some will do in the wild. Certain types of primula – and other genera besides – will commingle if left alone for long enough in the right patch. New varieties are made and their traits lie on their petals or their leaf shape or how well they put up with the cold. But they can also be traced through that scientific name. And while there are the grand and romantic stories of discovery – the hundreds of unfinished sketches and paintings made aboard James Cook's *Endeavour* by Sydney Parkinson, a Scottish artist employed to depict the botanical findings of Joseph Banks, before he died aged just twenty-five on the way home – I prefer the more humble tales. The hand-me-down plants which evade botanical classification and trade the race for conquest for something far kinder: generosity and enthusiasm. The Chinese money plant comes to mind. During the Seventies, the enquiry desks of Kew, Edinburgh Botanic Gardens and RHS Garden Wisley were increasingly being sent the same mysterious plant through the post. With flat, circular green leaves from spindly stems, it baffled the botanists. Terse responses landed in letterboxes: 'possibly a *Peperomia*', 'please send flowers next time', 'we do not identify sterile material'.

Here was a plant that was clearly being grown in homes around the country but wasn't on the books at any of these horticultural establishments. By 1978, a flowering specimen was sent to Kew, which led botanist Wessel Marais to trace it to a Chinese plant named in 1912 by German botanist Friedrich Diels as *Pilea peperomioides*. It had been collected by a plant hunter named George Forrest from a Chinese mountain region six years earlier. Forrest's collections were in storage in Edinburgh, but that still didn't explain how a plant used to growing in the cold, often springlike conditions of the Tsangshan range was popping up on the windowsills and tables of church bazaars in Middle England.

By the Eighties, public pleas for information were issued: an article was published in the *Sunday Telegraph* in 1983 – complete with an illustration of the plant – asking if anyone could explain where the one in their house had come from. When a Cornish family named the Sidebottoms replied saying their daughter had been given a pilea two decades earlier by their Norwegian au pair while on holiday in Norway, it took the hunt to Scandinavia. But the *Pilea peperomioides* even eluded Scandinavian botanists – they hadn't encountered it in their records and there was no recollection of it at the botanical gardens in Stockholm. Frustrated, Swedish botanist Dr Lars Kers appeared on Swedish television asking for information about this plant – and the network received 10,000 letters in response.

Sifting through these letters, which told of gifts and relatives and plants passed on, revealed an answer – and one removed from plant hunting or botanical science. Instead, the *Pilea peperomioides* had arrived in Norway in 1946 in the baggage of a missionary named Agnar Espegren, who had been sent home from the Hunan province in China two years earlier.

While making his way back to Norway, Espegren and his family spent a week in Yunnan before travelling to India. It was here that Espegren picked up a *Pilea peperomioides* – maybe from somewhere as simple as a market – and packed it up in a box for the long journey ahead. From Yunnan, the family travelled to Calcutta, and the plant went with them. When, two years later, the Espegren family arrived back in Norway, they did so with the plant.

Evidently, it thrived – until at least the mid Seventies all of the *Pilea peperomioides* in Northern Europe would have descended from this one little import. It's a plant that propagates happily: offsets (miniature versions of the plant) will appear in the soil. When they look big enough to survive in their own pot, you can uproot them. The Espegrens must have done this for some friends or neighbours, and they did the same – to the extent that one of the *Pilea peperomioides*'s monikers is the Pass-it-on Plant.

The process was taken to the UK in the Sixties, when the Sidebottom daughter brought one to Cornwall from Norway. And so for decades, homes all over Europe were learning how to look after and love a plant that botanists didn't know existed, giving it their own names, nurturing their curiosity without the structure of science.

The *Pilea peperomioides* made a voracious comeback recently. Perhaps in Scandinavia it never went away. But in the UK, it was one of the first unusual plants to prove that a renewed interest in plants was underway, especially among the millennial demographic.

As the obsession with minimalist Scandinavian interiors trends exploded in aspirational photographs on Pinterest and Instagram, the coin-shaped leaves of the *Pilea peperomioides* made their digital debut. Called the Chinese money plant or often just 'Pilea' – as with Tradescantia and Monstera, many of the varieties enjoying a new wave of popularity with new millennial gardeners have done so under an abridged version of their scientific name – the *Pilea peperomioides* became an object of desire, not least because it was frustratingly difficult to get hold of in traditional garden centres or nurseries.

Gradually, designer plant shops – created in response to the new, fashionable lust for houseplants – began to stock them in the mid 2010s and by 2017 it wasn't difficult to find a *Pilea peperomioides* in the traditional garden centres and flower markets – or in Ikea, for that matter.

But before the buyers caught on, people traded and sold their cuttings and offsets, advertising and selling them through websites such as Etsy or eBay as well as through Instagram. In fact, that was how I came to acquire mine: a funny little thing that has long rebelled against the plant's reputation for growing quickly. Before he became a good friend, a fellow south London gardener named Jack sent me a tiny *Pilea peperomioides* offset through the post. I had resisted buying one, and I don't think I could ever actually buy one in a shop, it wouldn't feel right. The entire history of these plants has been founded on inheritance. There is such – sometimes literal – kitchen sink beauty in the fact that when a plant has been happy enough on a cool, bright window-sill to propagate, the first instinct is to pass that offset along so somebody else can benefit too. So I nurture my little *Pilea peperomioides*, still so small some two years on, because when it grows big and strong enough to propagate, I will give that baby plant to somebody else.

•

There was an unspoken routine to visiting my parents. A simple but rarely changing formula of cramming onto the first off-peak train from Euston which, in early November, was aggressively heated. If you were lucky – and punctual – there would be a seat, but more often I'd carve a space on the floor as far away from the loos as possible and create a little nest around me from layers of coat and bag and Friday night fatigue. Dad would come and pick me, sometimes us if Hannah was coming too, up from the preferred 'secret' car park that avoided the taxi rank (one of the few things to truly incite rare rage in my father) and text us to say so as we bustled out of the station and into the damp air. Milton Keynes' grid system would melt into country roads I've known since toddlerhood; a few weeks away was enough to note how the seasons had changed in the hedgerows. And this time they were getting shivery, losing their leaves to damp piles that buffered up against the kerbstones. The Christmas tree farm, which never took its sign down anyway, had put a light on. To walk in the door was to hear the clarion call of my mother from the kitchen, and she'd come at us, all apron and arms, before bags were dropped. The alchemy of gin and tonic – always my dad's responsibility – would mark the beginning of a weekend in the countryside.

reset with every new persona I was throwing on; I learned to be a different person with my friends, my colleagues, Matt, my family. With the latter, I was always the baby of the family. In childhood, that meant I was the one who endlessly strove to play catch-up, to try to keep pace with their games. But as we grew up our patterns had changed. I'd carved different paths. Now, they had children and homes and people that made them into families. In comparison, I was both a lone entity with no clear future and something strangely dependent; the one my parents still worried about, warned to sleep more and work less and mind how I go.

London is too complex, too artificial to throw much exposure on a person. In the city, shapeshifting is encouraged and frequent. To go without seeing one another for months is normal, and if you've had a radical life change in between that's fine. But in the countryside things are slower. The pace of life is tracked in different pieces – a felling of trees, a new development of houses on a ring road, the decisions made by the Parish Council – and city trappings become garish. It is easier to shed them entirely for a few hours. Once those fancy townie ways have been left at the door, the freshness of the air is far better at revealing the person who carries them. Tales of the big city are shrunk by memories of the person you used to be. We went to see the fireworks in the nearest town and stood in the same park I illicitly spent my weekends in as a teenager, drinking cider and kissing boys. The things that are definitive in London – postcodes, jobs, friends and clothes – are meaningless here. I had long learned to scrub off those tell-tale parts of my city identity upon returning home; as a girl who had grown up in the country, it felt ridiculous to deny that history by wearing expensive trainers to walk across the fields. I'd slip into the same pair of hand-me-down wellies that had always stood by the back door, grab the same old novelty fleece that smells comfortingly of the outdoors and cupboard and then I would rummage in the drawer for a pair of gloves that come with a story idly told about whose they were and where they came from.

While other city-dwellers remark on the strange hush of small villages, I have always just sunk into the muteness, taking early nights and allowing those pure grey dawns to wake me. Days are divided by meals and between-meals, half-heartedly burning off the things that were recently eaten, rather than commitments or bus timetables. And in early November the gloom was creeping in. It was a fine place to witness those first frosts, the heavy moisture that clung to the air and the early swooping darkness, to make peace with the fact that autumn was turning from bright and crispy to grey and soggy and that this was a necessary passage to enable the life that would later unfold.

Ours is not a family that shows love in our language. We never learned to tell one another that we loved them, mostly because we were never taught. That announcement remains absent from our conversations and probably always will. It sounds like a cold admission, but the reality is more one of a gentle, considered distance. Voices were often raised in delight in the household I grew up in but rarely in anger. My father warned us against nurturing hate – especially when used in childlike petulance – and the extremes of emotion, such as sulking or hysteria, were hardly entertained. Love, I suppose, is such an extreme emotion, a science-defying part of domesticity, that this too became silenced. It's not that it wasn't present – it very much was – but it never occupied words. Rather, it would linger in the glasses of weak squash that we would be woken up with before school and the chocolate chips on top of the fairy cakes that waited for us when we got back home. It was passed down in the easy independence my parents granted me as a teenager and the wordless encouragement I was given to speak my mind. Pride and love rarely emerged from our parents' lips in as much as those nine letters, but sat quietly in the spaces our parents built for us to explore. I learned to adopt the traits I most admired in them: my mother's ability to feed whoever walked in the door well and seemingly effortlessly from the contents of the fridge; my father's abhorrence of waste and determination to remember the precise date that things happened. Others I'm still working on: his patience, her relentless curiosity.

And this love inhabited the bags of small, meaningful offerings it always proved near-impossible to return to London without. A few years before I'd cleared out my stuff as my parents left the last of the family homes, a final, pleasing sign of reaching a kind of adulthood. But other things made their way to London with me even so: finds from the giant regional car boot fair I always dragged my family to; linens and crockery and other unwanted domestic goods that had previously been loved by my parents and grandparents before times and fashions changed. Sometimes it would be a dog-eared, practical paperback my mum had found in a charity shop with me in mind, or a book I'd plucked off the shelf and wanted to keep reading. Other times it would be a rug bought on holiday that had finally ended its exile in the attic. And then there was the food – the margarine tubs full of cake and snacks not finished during the weekend, the packages of frozen stew. When I was between houses, especially, these were the most welcome things, the kind of gift that could dull the tedious, lonely ache of having to feed myself in an unfamiliar kitchen. My sister once told me, without envy or upset, that she didn't get the 'food parcels'. She'd long lived with the teenage sweetheart who would become her husband and always seemed so sorted and sufficient, that, as she put it, Mum probably figured she didn't need extra looking after. But I still get the food; even when I protest, it winds up snuck into my bag, a tinfoil discovery made upon reaching London. Perhaps it's because I'm the youngest; perhaps it's because they often see my life as one lived in a determined chaos – of work and projects and words.

That Guy Fawkes weekend visit was no different – I made the journey to the countryside with a small rucksack and returned to London struggling with bags. Food, yes, but different donations too. After years of gradually filling my home with thoughtfully chosen items, I had been left without a space to call my own and began to actively shun such pleasantries. Instead, I was sent back with plants. Hellebore (genus: *Helleborus*; variety: goodness knows) seedlings planted in the thick clay soil of my parents' garden, wrapped up in damp newspaper and put inside a paper bag. They were the progeny of the others that prettily littered the beds of my parents' garden in hues of pale pink, white and purple, themselves a gift from family friends who had moved to Devon. Hellebores, as my father says, are promiscuous. They hybridise, or cross-breed, themselves easily – a frustration for those gardeners who are precious about the varieties they have chosen, but a joy for those who enjoy being constantly surprised by what the earth turns up. The results can be positively rainbow-like, with unpredictable new colours and characteristics appearing only when the plant decides to flower. They take their time to do it, mind. Years can pass before seedlings finally flower, having grown strong and large and leafy. Once established, they will continue to brighten up a garden, pushing out unfussy, bold blooms when everything else in the garden is dying down.

I returned to London with another plant as well: a handful of soft, floppy leaves in a square plastic pot. My father presented me with it as I was about to leave the house. 'We found this in Grandpa's greenhouse,' he said, 'I don't know what it is, but I was wondering if you might want to look after it, see if you can make it do something.' My grandfather – my father's father – had died the previous spring, leaving a house full of sixty years of life and a garden and greenhouse containing the same amount of love to deal with. Dad spent the summer solidly working through both – my mum took the sweet pea shoots, got them to bloom, and others were adopted and planted in the garden. I took his clanking old steel watering can back to the balcony, where it proved a major upgrade from the milk bottles I had been using. But this plant had remained unhoused. Seventeen months after Grandpa had died, long after the house had been cleared and the Victorian greenhouse knocked down – the fancy new family who had moved in had no need for such a tumble-down, archaic construction – this little plant was still surviving. It went in the bag, came back with me.

That journey was not a nurturing thing. It was the first properly cold evening of autumn, the brisk winds funnelling down the train tracks as the minutes passed in the darkness. The deep wet of the clay had dampened the newspaper, which had spread to the paper bag. The weight proved too much for the wet mulch and the whole lot – hellebore seedlings, mystery plant, wrapped-up cake – crashed through the bottom onto the station platform. I scooped them up, carried them in my arms in a mess of raggedy paper and dirt. Hot train, stuffy bus, Sunday evening impatience, key in the door, home. I'd moved back into the flat a few days before, and could tell something wasn't right – the light didn't sift out of the windows as it usually did. Then I walked through the door to the heft of cigarette smoke and chatter of strange voices, the alien sight of the living room door being shut.

We'd long rented out the second bedroom in the flat; it helped with bills. But our previous flatmate had left in the weeks after the break-up and our attempts to find someone new were feeble and fumbled; we'd wound up with a childhood friend of Josh's who would pay the rent and wouldn't be in much. He too, however, was nursing a broken heart and lived out the trappings of partying and promiscuity within a space that we tried so hard to keep free and neutral of both for the sake of one another.

When I pushed open the door, I found the living room in a strange darkness with the furniture rearranged. Half a dozen people occupied the space that had previously been filled with a coffee table, books and magazines. Cans and pizza boxes littered the floor. Someone asked me – my arms still full of plant, coat still on, glasses steaming up – if I wanted a Deliveroo. The flat was a changing could crystallise with memories, the ones that cropped up in the flat or when I trod over familiar ground. Downstairs would have a party and I'd be taken back to the last one they had, where Josh and I stood on the balcony and looked on, ostensibly at the fireworks that lit up London's skies but increasingly at the fun going on below us. I desperately wanted to go, knock on the door, say we lived upstairs and ask if we could join. But back then I couldn't bring myself to raise the notion; it wasn't the kind of thing he'd want to do. And I realised that I felt relief, relief that I didn't have to hold back any more. That I could act upon my own mind, push into the more acute edges of life, the places that I used to claw at, hungry without realising why.

Such release arrived on a tidal wave of guilt. To reclaim space in my own life was one thing, but to become meaningful in someone else's was another. In many ways, Matt embodied this bold new existence, of freedom and spontaneity and a life positioned for pleasure. I often felt I had to show him my best side: he lived alone, kept wine in his fridge, ironed his sheets and was content in his own company and mine. I had none of these things and have always, ever since childhood, wallowed in panic when uncontrollable disasters strike. So I kept him away from it, painted a gloss over myself, turned the tiny tragedies of my days into irreverent quips and padded out the hollow joy I received from late nights and fashionable-sounding parties. For the most part, it worked. I convinced him, left myself to battle the unanswered questions when I was alone with my mind.

•

There was about a week before curiosity got the better of me and I got the mystery Grandpa plant identified, mostly because looking after something without knowing what it is was a fool's errand – a lesson I'd learned the hard way. There are apps you can get to do this. I spent most of my first months' gardening using them to identify the unknown beauties I'd bring home and to quell the fruitless Googling of words such as 'grey + feathery foliage plant' (artemisia!). The apps would proffer its scientific name and some vague tips on likes and dislikes, but it was enough to explore further, by looking the plant up in the library or, more often, on the internet. I'd glean more information, inevitably worry about the damage I had done and try to coax it on. Within a few hours I'd learned that the greenhouse survivor was a *Primula x pubescens 'Auricula'*, or a garden auricula among those who know. I'd never heard of it before, but the Google images were promising. They looked cute and cheerful, with different coloured petals and middles emerging from slim stems. The kind of gaudy cartoon flowers that filled children's TV in the Seventies and reminded me of the orange-and-brown graphic patterns that graced my mum's books of piano sheet music from the same era. A flower that conjured Puff the Magic Dragon and Zebedee all at once. When I texted my dad, he told me that they were one of his mother's favourites. I also learned that I was to ease up on watering and pull away the crisping yellow leaves – with any luck, it would flower in the spring, a couple of years after Grandpa had left it alone in the greenhouse.

With hindsight, it seems almost sacrilegious to associate auriculas with the latent hippy era of Middle England primary schools,

This was where the plants Clusius named *Auricula ursi* and *Auricula ursi II* were growing wild. Clusius was a Flemish botanist better known for his work on tulips, but he nonetheless managed to make these plants grow. From there, depending on which story you prefer, he either took them home to Leiden University or passed them on to his academic friends across Europe – as the pass-it-on *Pilea peperomioides* would be transferred centuries later. In fact, Clusius was such a fan of horticultural swapsies that he was actively against the conspicuous consumption of plants. 'To hell with all this selling,' he wrote in frustration in 1594.

We're yet to know the full details of how the auricula got from mid Europe to the UK, but it made its first appearance – as a Bear's Ear – in a British record in 1597, in the heaving tome that is John Gerard's *Herball*, which boasted five varieties. Flemish weavers have long been held responsible for bringing the auricula over to this little island in the sixteenth century while they were escaping religious persecution. It's a considerable myth in the auricula's story, both debunked (botanical historian Ruth Duthie dismissed it, saying the plants were too expensive for humble people to own) and clung to. The romance is somewhat irresistible, that idea of Protestant Huguenot refugees tucking auricula seeds into their luggage as they escaped the Low Countries and then northern France for London's Spitalfields, Norwich and Canterbury. Tiny bundles of hope and home to help settle into a strange new life. And while there is no paperwork to prove it, the story has stuck. There are records of Florists' Feasts – where growers and appreciators of flowers would meet to drink, compare and admire their blooms – taking place in Norwich in the 1630s.

But then, I know the slipperiness of history, how much of it evades time. My grandmother's ancestors were Huguenots. They settled in the fens of East Anglia to drain that expanse of sodden marshland. We can track them by their signatures and names – the Behaggs – appearing in the country from 1591 to 1655, but there's no concrete evidence that they were contracted to work on the Fens, or the 'Great Level', as it became known. These are the things that get sifted out like marsh water: a packet of seeds in a bag, a contract for a job, a settling and cross-breeding that results in a family story passed down the generations.

All that is left are the papers and the living results. Both have fascinated auricula fans. Because of their propensity to cross-breed, the number of varieties of auriculas is as difficult to pin down as their history. Within the first half of the seventeenth century the auricula had been cross-bred in such a frenzy that dozens of different varieties appeared. Stripes and doubles and colours and forms all proliferated in a wild display of horticultural creativity and control. These were recorded in drawings and paintings that later become objects of obsession for the auricula growers of the future after another grand migration – to the small gardens of the cities during the nineteenth century – saw such diversity dwindle. Modern growers have spent years trying to resurrect near-fantastical auriculas spotted in centuries-old paintings, a constant play of trial and error, science and surprise.

Just because some rare varieties vanished during the industrial revolution doesn't mean that auriculas wholly faded from favour. If anything, they proved resilient to all kinds of bigger trials, taking up different kinds of space in people's lives. As skilled tradespeople were pushed into factories, trading their craft for loom-wrangling, they indulged their love of flowers – especially auriculas – in Flower Societies. In 1822, there was such a club in every town and village in the manufacturing districts of the north; four years later, fifty auricula shows were held across the country. Although workers often only had one day off a week, the houses they were provided with came with gardens. These were small and often dark, but large enough to conjure a gardener's imagination – especially as gardening enjoyed a boon during the nineteenth century anyway. And it was here in these smog-filled stone plots that the auriculas continued to flourish, possibly even relishing the chill and the shelter provided by the meagre conditions. Auriculas even crop up in the miners' cottages and working-class gardens of D.H. Lawrence's novels. Sniffed at by the horticultural establishment as 'a poor man's flower', they brought delicate joy to difficult lives.

Even with little time and space, growers were fettling with auriculas. The second half of the Victorian era saw an increase in black appearing in auricula flowers – it's thought people bred it in to reflect the fashion of the time, which in turn had been influenced by Queen Victoria's mourning of her husband Albert, who died in 1861. People have always been drawn to their fussier requirements. Auriculas don't like to be left in soggy soil or extreme sunshine. They prefer shade and cover. For the show varieties, whose petals are covered in the fine white dust known as farina, a drop of rain can spoil a pretty flower. All of which led to the invention of the auricula theatre in the eighteenth century, in which the potted plant would sit in rows, protected from the elements and on display for all to see.

Some gardeners still meticulously maintain them now, but it's a level of care I can't get on board with – these are plants that smother mountainlands, happily cross-breeding with one another to make fantastical-looking flowers. To me, auricula theatres represented an old-fashioned, fussy form of gardening. But that was all right, I already had one on the balcony: shaded from the rain by the roof, from the sun by the walls and kept dry and cold on the concrete. The auricula didn't flower in spring, but I stayed patient. By midsummer, tight buds on skinny stalks started to emerge from the foliage. It flowered in August, a heartstopping display of apricot and raspberry and peach; a Barratt Fruit Salad of a flower bursting into bloom out of time with the season and out of time with its generation. A product of love passed on but surviving nonetheless.

DECEMBER

'I LIKE YOU,' HE SAID.

I'd been feeling the weight of it coming, like hearing the rumble of a goods train heft its way down the track. And I liked him too: liked the length of his eyelashes and how lightly he carried himself, with ready humour and plenty of kindness. I liked how he looked when he was deep in thought, furrowed brow over high cheekbones; how his silhouette caught the light in the cinema. I liked that we could lose track of time by talking; that our conversations would stretch and lengthen; that we would rather harvest one another's curiosity than do anything else. I liked waking up and finding him there, and the gentle steam of the cups of tea he would make me while it was still dark.

But it made my stomach lurch. The strength of feeling between us was large and unwieldy. As autumn drew cold and dank I'd told myself and my friends – even him – that this was just a fling. I'd brush it off, use the concept of 'cuffing', the half-serious notion that as the year draws to a close our commitment-phobic generation couple up to cuddle away the darker months together before breaking up in the spring, renewed and single and free. I liked to compartmentalise my life, to keep work and home and friends and family in their separate pockets. Matt occupied another one entirely, and I tried to keep it small and tidy. To let it gain substance and strength would be to defy all that I had been advised by the books and the television shows. *Eat, Pray, Love* and *Sex and the City* had all dictated that I should sandwich my life with periods of meaningful solitude; that there should be time for me to indulge in myself, to weather my feelings and pour all that leftover love back into me. And I'd spectacularly failed at that; I had fallen into a new something – not relationship, not yet – while so much of me was still fascinated by what had gone wrong with my previous one. To voice how much I felt for him would make everything hard and real, unravel the artifice I was feebly trying to weave around it all.

The end of the first weekend of Advent was nearing; several hours had passed since a mid-afternoon sunset. Eventually, I looked back at him, met his eye. 'You have to be patient with me,' I replied, slowly, like a sigh.

'I know,' he said, just as softly, with no heaviness at all.

•

Ours is the first modern generation to share houses well beyond our twenties; in London, it's often the only affordable way to live in a decent location. As we grow up, things improve: the bathrooms become less grimy, people hire cleaners and divvy up the cost just like any other bill. But we still share, make families from our flatmates, sometimes find ourselves longing for more space or privacy. Partners and tentative steps on the housing ladder come along and the groups change, like line-ups in a pop band. We live with individual kitchen cupboards and our own shelves in the fridge, towels on racks and low-lying irritations side by side with the conviviality. Knowing that to come home is to find friendship and chatter.

I'd stopped house-sharing in my mid twenties, when Josh and I moved into the flat. But now that I was out of it, I was back to communal kitchens and subletting. It had been a while and I looked forward to it, keen for a sense of community I had realised I'd lacked. I thought that the solution to the gnawing loneliness that had emerged from our beautiful grown-up and distant existence might lie in finding a group of flatmates, the kind of people who become bridesmaids at weddings and best friends on holidays. I idealised it, cherry-picking gilded moments from my memories. I imagined sitting around a well-worn kitchen table drinking red wine, listening to music and meeting new people's old friends. I wanted some more mess, I craved a kind of commune, the unique intimacy that comes from being flatmates who actually like each other, spend time with one another. A ready-made belonging in a world that felt increasingly segmented.

I found a room in a house in Homerton, a stone's throw from where my school friend Anna lived. The doorstep was tiled like a checkerboard, and as I walked over it and through the red front door my glasses steamed up, acclimatising to show a handful of women clad in black and idly bustling around the kitchen – cooking, eating, reading, talking. As many bikes were stacked up in the gap between the fridge and the kitchen door. The lighting was bad, the windows throbbed with condensation, the remnants of the interior design dreams of a long-ago custodian clung to the wall – homespun brickwork and early Nineties cabinets in an orange-brown sprawl. A large *Pilea peperomioides* sat on the kitchen counter, little offsets appearing at the crown. It was a house both unfamiliar and instinctively homely, recalling the kitchens of other houses I had shared, of other houses my friends had. As they nattered and introduced themselves, I felt my body exhale a little. I could settle here, certainly for a few weeks. Longer, if I needed.

The room was simple and tidy; a rhipsalis dangled over the desk. Next to the window was squeezed a bed that I would only ever sleep in alone. It had been left neatly made beneath a flaccid hot water bottle. Above the headboard, bizarrely, was the same poster I had hung in the rooms I'd inhabited for the past decade – a whitewashed collage of photographs and sketches that I'd pulled out of a French art magazine as a teenager. How the thing had got there still baffles me, but it gave a click of recognition. That this was beyond comforting; it was correct.

There was a garden. A wall as old and red-bricked as the house kept it separate from the alleyway at the back, a wooden gate in the middle. Beds surrounded a paved patio. My favourite bit was the damp shady side-return that had seized neglect to become a fertile home for ferns and ivy which had spread around an old wooden stepladder. When I was awake late enough for it to be daylight, I'd look out at it from the kitchen table and think about how that could have been a scene from any decade before. The scruffy bit of the Secret Garden, the left-behinds of the pampered lawns. The table was dressed with a single white pom-pom of a dahlia in a jar. Another seat at a table with a garden view. I liked it there.

One of my new flatmates would work in the garden. She knew her stuff, too. I came home one night to find an enormous mass of tubers sitting on paper by the back door: uprooted dahlias, dark clods covering their creamy hulk, strange tendrils of root emerging at every angle. The paper was neatly labelled in green ink: 'I'm drying out. I'll move soon, soz'. A few days later they had vanished – I'm not sure where – to stay dry and cool over the winter, avoiding the frost before having another season in the soil next summer.

She said that her mother was a keen gardener, that she had learned a bit from her, but her enjoyment of it was familiar to me: a quiet, near-everyday thing that brought a pleasant escape from the rigour of city life, even in the depths of November. It made me realise that there were probably lots of us reaching for spades and drinking in what we could of our parents' and grand-parents' knowledge to better plunge into what soil we had access to, just to feel a bit better, a bit calmer.

Nobody knew, though, if those tubers would go back into the soil the next spring. The house was on the cusp of change; discussions knocked, gently but persistently, at bedroom doors and at vague futures. Leases were coming to an end, there were new lives to be embarked upon. The gardening may have happened but it was squeezed in on cold Saturday mornings, while I dragged myself out to work weekend shifts. The other women were equally busy – most juggled two or three jobs simultaneously, split themselves between practices, teaching and writing and their own creative efforts. We bumped into one another in the hallway, grasped at greeting and politeness, but there was rarely time to have a serious conversation between work and social lives and the ever-increasing tiredness. As the weeks wore on I realised that just because people bustled around in the kitchen at the same time, it didn't mean they were necessarily eating together – they simply occupied that same space. The joint houseshare WhatsApp groups dealt with life admin more than love. We had so many options; we could make our careers out of cocktails of different trades; we could spend a few weeks living in one part of a city and then move to a new neighbourhood. On one level this was freedom, unshackled from the strictures of permanence – mortgages and jobs-for-life – that our parents spent their young adulthood chained by. But the reality was something far more draining: a sense that we were always working without a clear reward in sight; never knowing when we might have to find a new home. These situations couldn't foster any sense of community, we were pulled apart in too many directions to cling to anything more permanent.

I wanted to go out in the garden, found myself staring out through the glass of the back door as the kettle boiled, looking to see if the key was around. But I never found the courage – nor the overlapping time – to ask where it lay. It felt like it was someone else's land, certainly not mine to pick around in, inspect the things that were growing, the projects that were silently unfolding. This should have been a space for me to ask to join; instead, I found myself shut out. I couldn't just drop into other people's lives for a few weeks and expect them to take me in; we already fill our hours with so much. It was easier to balance life when we lived it in boxes; we seek community and connection online, through our digital selves and the friends we make at work, kept separate from the friends we know from home. I would walk around the corner, to the estate where Anna lived, and have cups of tea with her there instead.

•

I'd emailed the people who ran Brockwell Park's Community Greenhouses the day after stumbling across them with Matt. It hadn't taken long for me to turn up, the minute the gates opened, for my first volunteer shift. The city was in the midst of a November cold snap. I was set to work with weeding couch (pronounced 'cooch') grass out of a large plot that would become a wildflower meadow. The gardens at Brockwell Park Community Greenhouses are kept organically, and so the way we rid the earth of this most adamant of weeds was to root it out. I put on gardening gloves – mostly for the cold, I was happy enough with the soil – and set to unearthing those bright white roots, the innocuous-looking leaves that grew from them. Every task was filled with novelty: until that morning I had not even heard of couch grass. I gulped down the educational scraps about roots and nitrogen thrown out by the head gardener. There were other things that I could never have done on the balcony: potting up aquatic plants, moving bindweed into a skip. I left covered in compost and glee, my arms and back aching from the strange newness of using tools with long handles. My fingers felt pinched with the cold and the grit; I wanted to toughen them up. This was gardening but with a bigger purpose, tending a space that wasn't attached to me. By being there, I was removing myself from the notion of home, of doing it to improve my own little abode. It wasn't that there was anything wrong with growing on my balcony – I wouldn't have started if it wasn't for the promise I saw in that space. But the community gardens showed me that gardening didn't have to be something that happened in a back yard, in a private place. That the liberation I had found in London's parks, the strength I had seen in the stubborn, self-seeded plants that had grown in its rubble and railways, could be found and tended and encouraged by coming to this spared space.

The greenhouses offer room to pause and retreat. There are two of them in the community garden plot – one easily the length of a double-decker and the other just shy – and all containing the pleasing atmosphere of rag-tag industry. Seedlings line up in pots, homemade labels bear warnings of overwatering and enormous tropical plants push against glass ceilings. Each boasts a different climate (the lower greenhouse is kept warmer), although most business took place in the upper greenhouse, where lessons and little ones and curious locals would pop in to tentatively look or sometimes buy the things that had been grown. Seedlings, different varieties depending on the time of year, grew on long metal staging beneath a grapevine-covered ceiling. At the back, two doors led to a smaller, quieter room where *Pelargoniums* and succulents thrived. It had the air of a sacred space, one for those who nurtured these more special, more fussy plants. Handwritten signs warned 'only water on Sundays!'

These glasshouses were custom-built in the early Eighties, but they sit on a history of produce. People had been growing for 200 years on the plot where they stood: some of the acres contained within those brick walls had originally been a kitchen garden for Brockwell Hall, the elegant house on the crest of the hill that gently rises above the gardens. The manor was built in 1812 by a man named John Blades, who had made his fortune in the city as a glass merchant. Blades set up a chandelier shop in Ludgate Hill in 1783 and had soon wooed the great and the good with his designs, exporting chandeliers and glassware to the British aristocracy in India before cutting glass for King George III. He was becoming an old man when he bought the land that became Brockwell Park, instructing architect David Riddel Roper to crown the stretch of gently undulating parkland – then considered part of Brixton, Surrey; now considered part of Brixton, London – with the vaguely Grecian building that has now been a café for longer than Blades was alive. The estate was accessorised with lodges and roads and fences designed by Blades' sometime collaborator, John Buonarotti Papworth.

Blades' land made a charming vista for the inhabitants of Poets' Corner, a smart new development of houses on streets that stood adjacent to Brockwell Hall's grounds. A couple of decades after they had been built, my great-great-grandfather Alfred Behagg had moved into a smart family house: number 12, Shakespeare Road. The Behaggs missed the opening of Brockwell Park by London's City Council in 1892. The same year, my family moved back to the Huntingdonshire countryside, returning to the fenlands where their ancestors had first arrived. But as I walk the park I like to think of them: Alfred, his wife Louisa, their children and the assorted siblings-in-law that filled that Herne Hill house. Maybe Isabelle, Louisa's sister, walked there with a sweetheart; she never married but the Valentine's cards she received are still held within my family. Perhaps, on Sundays, the family maid would have spent her free time there, inhaling deeply from green lungs.

The park had been transformed for the public, landscaped by J.J. Sexby, the chief officer of the London City Council's parks department, and filled with the furniture of late-Victorian public parks. Ornamental ponds were dug and bandstands were raised; frilly carpets of flowers filled the beds around the Hall. A strange clocktower was uprooted from a fairytale and placed in the middle of a path. The walled garden was partially released from service, kitchen beds dug up and paved over, rose bushes planted and promenades installed. Sexby tweaked and topiaried, making an old English walled garden in the process. A well stood within a twee wooden structure, a leaning mulberry tree was left alone and remains in place – people have been drawn to it ever since, mystified by its age and significance. Photographs of the walled garden from 1904 show women in hats, gently holding their floor-length skirts under trees that had shed their leaves for the winter, and now it looks much the same. It is quietest on Sunday mornings in the winter, where life is gathering steam beneath the soil.

It wasn't all for show. Some of the kitchen garden retained its productivity; Lambeth Council grew bedding plants there that would go on to cover the whole borough with colour. Now, far less of the community garden is used for profit, but small trades happen with the produce that is grown there. A few coins for as I threw myself into summer had suddenly arrived and could be motivated by the smallest of injustices. I took his occasional unthinking – not texting me back, or forgetting plans that had been made – and turned them into a source of fury that I refused to let him see. Instead I would seethe, become agitated and wait for such tiny storms to blow over, when I would swallow everything down. I wanted to be both adored by him and kept at a distance. I had been so blindsided by the rush of a new fling – of being seen anew by a stranger, of having them want me when I felt most unwanted, of being able to exist as a polished version of myself – that I was unable to understand what happened when it turned into something more steady. Most of all, I was convinced it would end before it had even begun, that I would be left feeling tricked again. That the minute I let myself ease into comfort, into softness, it would be snatched from under me. Sometimes he called me his girlfriend and it felt like a hot blade was brushing against my skin, catching only the finest hairs. A dangerous, thrilling near-miss. I wanted to be with him; I felt myself pulled to him, to his home, to his life; a need readily and gratifyingly entertained, but I flattened those feelings with denial.

One slow, near-winter weekend was too full of meals cooked on his stove, dishes washed in his sink, showers taken in his bathroom. I felt muted – not by any recalcitrance on Matt's part, but by a familiarity that shocked and scared me. This most gentle of challenges – to lose unspooling hours in Matt's home, to pick up a few groceries while he worked – arrived like an assault. I knew I was playing a part; my personality felt conflicted, as if it consisted simply of geometric shapes that didn't quite fit together.

the dark brickwork, purple salvias grasped onto their last blooms. Doors and windows supported the skeletons of vines and trees that, just weeks earlier, would have painted burning gashes across the buildings. Now some yellow leaves prevailed before littering the crooked streets below. Pots held the remnants of weary hollyhocks, brave *Pelargoniums* flowered in boxes attached to the bridge rails, jostling for space alongside the handlebars of well-loved bikes. Some shopfronts offered planting better suited to the season: heuchera, ivy and euphorbia foliage jostling for space in a flash of green. By mid December, the Jordaan was part-festive forest, too. Man-size Christmas trees – pushed there on bikes – were propped between houses and covered in white lights. How tasteful, how handsome, offering a Christmas that was infectious in its subtlety; a Christmas that recalled those of decades, centuries before. A bunch of mistletoe tied with a red ribbon hung on a black door.

That dislocation stretched further with nightfall. We had napped through sundown, so emerged into that inky beauty without seeing how it had arrived. And such beauty, it held me like a fist. The moisture in the air had cooled, settled on the stones beneath, and the light from the elegant lamp posts became a glimmer, palely bounding off canal water and pavement. The clarity of the day had become a fog by night, and this smeared with the lamplight and water like an oil painting, removing the trappings of modernity and placing us somewhere out of time. The canals swallow sound, but there wasn't much to begin with: the cars were all parked, only bicycles passed by. The still of it all caused us to drop our voices, draw a little closer to one another.

It was a charming backdrop for romance. And there we were,

two people too nervous to fully express their feelings. Mine had been a right jumble, anyway. With Matt, when I was in his presence, there were no questions. Answers didn't appear either, but without the constant curious patter in my brain they weren't warranted. To be with him felt increasingly simple and safe. I was no longer wary of knitting my fingers with his in public, or to click into the space between his arm and his chest when we were alone. These collisions created a space of mute security, an incubation vital for something precious to grow. Something contained, in that I knew that those hours offered just a temporary distance from my problems beyond, but visceral nevertheless.

There was a reality to us that was undeniable. As he appeared with a handful of kisses, breakfast bought from a fashionable café over the canal and whisked up four flights of stairs, I was beginning to accept that. I was learning, too, how our habits could jar with one another: my approach was swift and practical, often too hasty and driven by the need for solution. He meandered and hesitated, tried to fit too much into a day and didn't worry about the fall-out. That morning, he had to write. I had been a journalist for years but I'd never sat and watched somebody else pull a piece together before. There was an almost unbearable intimacy to it, and it transpired that I was an impatient observer. I found his frustration difficult to witness. I felt cooped up by having my time partitioned by somebody else. And so I took myself out, wanting the clear skies outside from which to breathe better. The same familiar craving for escape – the one that kept sending me out onto the balcony, that drove me to explore London's parks when I was steeped in loneliness, to the community gardens a plant originally stolen by an East India Company employee in Yemen) came from Java to Hortus in the late 1680s, whereupon they were cosseted in a hothouse. Surrounded by these fake tropics in the damp fug of a northern European city, they grew and in turn set seed. From there, the story goes, offspring of this plant were given to Louis XIV in 1714, who planted trees in his gardens. Once in the hands of the French, coffee plants were taken to their colonies in the Caribbean and on to Brazil, leading to the explosion in European coffee drinking in the early eighteenth century. Curiosity among the Hortus staff had led them to roast and brew the seeds to make coffee some years earlier, however. That drink still fuels the city, from the modish, minimalist workspaces to the brown cafés, or *bruine kroegen*, which occupy the street corners with wood panelling imbued by decades of cigarette smoke. Matt and I spent an evening in one, accepted outsiders sitting at a table in the corner and drinking alongside locals in a room where it felt like the clocks had stopped.

Hortus kept growing beyond those first heady decades of discovery. The palm house – a soaring, steampunk-looking thing of yellow brick and girder – was built in 1912 purely as a bargaining chip to keep the incumbent Hortus professor Hugo de Vries in Amsterdam after he'd been offered a job in New York. By the late Eighties, Hortus was on the brink of bankruptcy, but the community stepped in to shift its fortunes. Hundreds of years after they brought about its opening, the City Council once again began to cover costs, and in 1993 a new glasshouse was built, one that created the three different worlds – of subtropics, desert and tropics – under a lumpy, transparent roof on the edge of a chilly canal.

And this was where Matt and I wound up, staring at that blue-skied mid-December afternoon through the haze of glass mottled by condensation and the moss that fertile air brings. We sat on a bench in the region that claimed to be South Africa or Australia, surrounded by the thin needle-like foliage of plants developed to retain moisture in dry air, the soft sprays of willow bottlebrush and the strenuously elegant boughs of tree ferns. Ascending the spiral staircase took us to a runway where we could look over treetops and play with the place's strange boundaries, of control and nature, life and fiction. A clump of wiry maidenhair vine nestled above an air vent quite happily; a waterfall trickled from a pipe; a gardener's anorak hung over a steel rail, tickled by the tendrils of tillandsia. I identified what I could by the Latin names, while he stuck to pronouncing the Dutch ones in a bad attempt at a local accent. Yesterday-today-and-tomorrow, so named because its flowers change colour as they die, kept its English moniker.

The place was deserted; botanic gardens often are in winter. The canal ducks took free reign over the leaf-covered beds outside, signs hinting at the spring that was to come – *Galanthus*, or *sneeuwklokje* (snowdrops), crocus, hyacinth, the fattening buds on the rhododendrons. And we roamed it too, ambling over to the Palm House, which is full with the plants that spend the warmer months in the conservatory. A temporary jungle more than a century old. There were some cabinets in one of the wings that held dusty relics from those days of seventeenth-century exploration, as opposed to the living ones they were surrounded by. The air – drier here than in the tropical house, which was damp and heavy with contained life – felt trapped in time. A neatly sprawling Wood's cycad stood in a wooden barrel nearby. It bore the label 'extinct', having been brought to Amsterdam in 1855. This was a male plant, which Hortus had managed to propagate from suckers. No female plants have been found since. It was a green dinosaur living alone.

It was the Dutch who created some of the first proto-greenhouses, as early as the 1680s. From there, the concept was snaffled by British aristocrats (Queen Mary II had one built at Hampton Court) and horticulturists – a couple had cropped up at Chelsea Physic Garden within fifty years. They were built, as with the rather dingy orangeries before them, to grow exotic fruit such as pineapples and tropical plants shipped in from the colonies. But they were also buildings that dripped with affluence. As the eighteenth century shifted into the nineteenth, glasshouses gained ever more glass (their first iterations, known as hothouses, were more like huts with windows) and became major status symbols. A fierce rivalry broke out between two schools of greenhouse thought – iron frame versus wood – as estates across the country were bedecked in 'seemingly airy and unsubstantial forms', according to an 1837 edition of the *Gardener's Magazine*.

By the time Victoria was on the throne, some 200 greenhouses had cropped up from one manufacturer alone and 'a mania for conservatories' had ensued, as one observer noted at the time, and 'spread contagiously among all the richer classes. They are attached to all the more pretentious houses.' What had begun with plants became, in some cases, more about parties: the conservatory at The Grange in Northington exemplified the pinnacle of greenhousery in 1824, with curved roofs and climbing plants trained up columns, but it wound up becoming a ballroom. By the 1840s, a combination of new railways divvying up the country and an unrelated end to a punishing tax on glass saw greenhouses delivered up and down the country, encouraging the middle classes to create tiny, fertile worlds in their own gardens.

The losses of the First World War put an end to the grand glasshouses of country estates; gardeners were killed on the battlefields, and there was no longer the money to fund the upkeep of these lavish transparent worlds. The glasshouses were left to be taken over by the plants, panes shattering under tenacious, hardy vines. Some, such as the fernery at Ascog, were rescued by regeneration missions in the Nineties. Others, like the eerie, domed glasshouse at Hilton Hall, stay skeletally standing, half reclaimed by nature, half a vestige of man's one-time ambition.

Millennials revived the Victorian fascination with glasshouses. Not by acquiring them in the physical sense but through the narrow window offered by our smartphone screens, where a whole world of glass and condensating green can be accessed. The swooning curves of Kew's palm house; the vines that smother the spiral staircases in Edinburgh's Botanic Gardens. The way that *Monstera deliciosa* become dwarfed by the brutal concrete heft of the Barbican Conservatory, itself only built to cover the enormous fly shaft attached to the complex's theatre. The ethereal clouds that billow through greenery in Singapore's sky garden and the narrow, smothered tropical corridor of Chelsea Physic Garden.

Beyond the public glasshouses lie the less accessible ones. Like rare ferns in the nineteenth century, the elusiveness of these has made them even more desirable. The small conservatory of the South London Botanical Institute is beloved because of how rarely it opens. The barely contained wildness of a greenhouse full of cacti owned by an elderly man named Richard in Robin Hood's Bay in Yorkshire has gained its own kind of fame. Hand-painted signs on the roadside advertise 'cacti for sale' and explain, inside the greenhouse, that 'all money raised by the sale of these plants goes to help children in the third world'.

While these spaces may be oft-overlooked in real life, their digital representations have inspired wanderlust and relief in millennials whose greatest stability lies in a twenty-four-month phone contract. I hungered for a place of my own just like thousands of others my age, whether that was the fantastical walled garden I sometimes daydreamed about or something as simple as a residence for more than a few weeks. The notion of a permanent home had become a fallacy, something not to be relied upon. A garden was an even more unthinkable privilege and that placed greenhouses – so long a trapping of the retired, suburban or merely quiet – in a position of strange new luxury. Admittedly, they looked good on Instagram: the photographer creators of Haarkon rapidly amassed an Instagram following of hundreds of thousands by posting beautiful images of greenhouses they had hunted out on their adventures around the world. Hashtags grouped the shots together – '#ihavethisthingwithglasshouses', '#greenhousehunter'. Caro Langton and Rose Ray, creators of Studio Ro-Co, had their online success bolstered by photographs of their House of Plants, the name given to a rundown coach house in Hampstead that was blessed with a conservatory the size of a studio flat. When I visited Clark Moorten, custodian of the Moorten Botanical Garden and Cactarium in Palm Springs that was founded as a 'cactus museum' by his father in the Thirties, he mused that his humble greenhouse – more of a bunker, really, covered in corrugated plastic – 'must be the most Instagrammed greenhouse in the world'. Young, made-up women in kaftans wielded SLRs behind him, the better to contrast their glamour against cacti that had grown over the decades to cover the ramshackle staging.

This unlikely rise in greenhouse popularity thrummed with something I had felt since childhood. The physical release I find in gardening, in growing things, has many origins but my grandfather's greenhouses are among them. That giddy confusion between being both indoors and out, of making magic from nature and the tiny triumphs and failures such spell-making contains. The lost hours and careful minutes, the littleness of seeds and trays and pots and the grand scale of ambition – to grow grapes and tropical plants in Sussex and Reading! – all collided to make these spaces special. Square metres of quiet and repose for the sheer pleasure and challenge of growing things.

Grandpa's house was partially flanked by glass; the greenhouse was tacked onto what he called the scullery, but the front door was surrounded by a porch of intricate panels. I never really noticed it when we visited – the double doors were always open, and we'd bustle in to knock on the front door and await the sound of him shuffling down the hall some minutes later. I was surprised to notice it some years later, when I looked up the listing online and saw his house in the estate agent's photographs.

The light coming through the panes was green, bouncing off the magnolia tree and overgrown hedges of the front garden. This porch, an elegant Victorian creation, used to woo a new, anonymous owner: I felt like I was seeing it for the first time.

I wasn't yet twenty when I sought out Grandpa's greenhouse for a university project, trying to satisfy the urge to retreat into my origins mere months after moving away from home. But even scrambling over the piles of plastic pots and bags of compost – the space had grown messy with Grandpa's tenth decade – bestowed me with a kind of full-body sigh, a release that comes from occupying a space of concentrated growth. A week after he died, I found myself back in that greenhouse, taking in the smell of the spilled earth, too overcome to notice that all the plants were still growing.

Because things do continue to grow. Plants exist to live just like we do, in spite of bad days and confines. In spite of the punishing controls that we suffer under and that we put on ourselves.

Away from London, in a charming continental bubble of another city where time seemed to slip and slide as readily as a bicycle tyre on wet cobblestones, the pressures I'd been putting on what was happening between Matt and me eased. The performance slowed. I couldn't pretend to be a more perfect version of myself for three days straight; that woman shifted, became somebody who got crotchety when we waited till eleven p.m. for dinner and became better at expressing her own mind.

In this city were weren't marking familiar ground or working over landscapes that had been formed for or with other people, but discovering – even building – a new one of our own. Amid dozens of strangers, in a portion of time where I could just be alongside Matt and out of context of the complex trappings of home, we could finally give space and time to that which lay between us. It was so freeing, as if the vacuum seal had puckered and let in the smallest gasp of outside air. Much-needed oxygen, better to clear the fug of my fears and allow us both to breathe beyond the boundaries.

•

The London we flew over was crisp with rare winter sunlight. I clocked the ridges of the Thames Barrier first, white in the bright of midday, but the rest of the city soon unfolded. A mass of pale brown and red gobbling the green-grey reaches of Essex, the Thames a leaden snake through it all. We lowered as we neared the airport: the dated thrust of Canary Wharf; the upturned spoon of the Millennium Dome; the Old Royal Naval College marking Greenwich like a doll's house; over the sprawl of Deptford until I was presented with an aerial view of my local supermarket, and the crawl of Dog Kennel Hill; Ruskin Park and then the unmistakable outline of the block of flats where I, technically, still lived. It pulled a strange grief in my stomach, for here was a new vision of a landmark that I did not know how to process. Mine in deed but not really word and increasingly not sentiment, a place I was actively learning to un-love for self-protection. It rapidly put my dislocation into context: returning from a holiday with a new lover only to be surprised with a view of where the previous one was sleeping. I swallowed it down, carried on looking. We marvelled at the gleam of the sunlight on the top of the Shard, so tall it seemed within touching distance of the plane wing. Another glass house, another construction to house whim and fantasy and status.

•

Christmas Day began with the strains of Wham!'s 'Last Christmas', chugging through the speakers of the Uber that was driving me back to the flat after a Christmas Eve dinner across town. London was deserted; the annual pilgrimage to the suburbs and the countryside completed several hours, days ago. I was not among them. This year, Hannah was hosting in Lewisham, a goose and a six-week-old baby to make Christmas in a new home. I woke up alone, to the flat grey light of a well-worn pillowcase. The city lay quiet, and I cycled through its ramparts, puffing over the ducking and diving rows of terraced houses that constitute south-east London's furthest reaches. Past a man being put sullenly in a riot van. Then to her front door, wrangling the bike against a pavement railing on a day when even thieves wouldn't stir. The day unfolded in cheer and wine and Pixar films, cracker hats discarded on the floorboards. I woke up on the sofa in the early hours of Boxing Day to read that George Michael had died.

The news struck me unexpectedly. My day job is to write about pop stars, and I have grown used to their unexpected demise: the initial shock, the outpouring of tragedy, figuring out how to sail a small ship of meaning across it all and come out with a way of summing up a life. But I was saddened by it, how quickly his life had been extinguished, how I'd heard his voice ring out just hours before he died. On Boxing Day, I cycled back home via One Tree Hill, a nub of a park surrounded by houses, hard to get to on a bus. From there, you can capture the city – it has far better views than the more famous spots. Further away than Primrose Hill and less busy, so you can watch the houses grow grander, the tower blocks rise until it merges into a tight knot of grey, dotted with postcard landscapes. This was the same view as the one that the flat offered – inferior to it, even. I had often thought that when people spoke about the fashionable rooftop bar in Peckham or the pub garden next to the estate, and marvelled at what they could see, that I had a private window to it, and the comforts of home, so why leave? Why entertain something diluted beyond those walls?

But I chose to be here, looking at that familiar landscape from a new space. Soaking it in from a park bench instead. Taking in its outsideness, the breeze against my skin, the passing families walking dogs, the placement of me in this space, a brief borrowing of it. The chill seeped into my bones as the daylight drifted into ash, and I watched my breath cloud in and out as 'Careless Whisper' seeped through my earbuds. My whole body ached with sadness, caught between a desire for no company other than my own and a profound loneliness, at being on top of a hill, alone, on Boxing Day, listening to the voice of a dead man. Christmas had been reduced to its bare minimum, cut short and strange. I felt out of place with tradition, out of sync with the festivities. I waited until it grew dark, then made my way back home. The next day, I returned to the office and revelled in the silence and solitude of it; there was a freedom in the oddity of being at work as everybody else made road trips to visit family or ate Quality Street on the sofa. I listened to 'Faith' obsessively for the next week.

Christmas faded from view; I worked it away, pushed it away, determined to get life back to a sense of normality while the city was stilled. I clung to the tipping of a new year as a kind of salvation for the first time since adolescence, when the end of December had been something potent and meaningful, rather than a time riddled with expense that never quite lived up to expectation.

Matt and I had been at a party in north-west London, the terraced house of a newly famous person and her rich husband, filled with people whose faces I knew from the telly. It was a New Year's Eve Eve celebration, something to start off a long weekend that I hoped would banish the previous year in a wild final purge. It was also the first thing we'd been to as a pair. I spent much of it feeling deeply uncomfortable; not knowing what Matt would introduce me as, knowing that it would be strange regardless. He found friends and colleagues to catch up with and their names rattled through my brain like glitter. I had got the dress code wrong, felt I looked too frumpy amid the frocks and fur coats. People were kind and friendly, increasingly so as the night wore on, but I was lost in it all, didn't know which face to wear, which person to be.

As midnight tipped into the last day of the year, though, we found ourselves in the crowded, sweaty living room. Giddy on fizzy wine and the first few cigarettes in a while, he pulled bizarre, silly moves and I felt so relieved, so comfortable in the ease of it all, in the gleeful bubble we had blown around ourselves. We sank into the music, shuffling and laughing. It was as if I was lit up by it, losing track of weariness or want. When we reached Warren Street, we were the only two people on the platform; it must have been nearly four a.m. He stood opposite me as I tucked myself into the tiled alcove, and, with the confidence and mischief of someone who already knows the answer, asked me if I loved him. 'Sometimes,' I replied. 'Sometimes, I think I do.' And it was true; I had felt the words rising out of me in recent weeks, when we were wrapped up in one another, or I pushed back his hair. When he did something sweet and unspoken, like buying in the childish cereal I liked to eat and he didn't care for. I felt it grow in me, a new love.

And then it was out, hovering in the stark, fluorescent lighting of the Tube. It felt so precious, so raw and so new, and I was baffled by it, this love. How it had been incubated, where it had come from. But it was there, and he loved me too.

JANUARY

I WATCH THE SHORTER DAYS meticulously. The longer ones catch us by surprise, greeting us with a bright dawn or tricking us into thinking the evening hours belong to the day. But there is no such absent-mindedness with the short days, which take their time to wake up and then, docile, fall back to sleep again, rushing us back into houses and pubs, ending walks and outdoors things. During the shorter days, I barely see the light. I prefer to start my work when it is quiet, early in the day, and dawn crests while I sit in the office. Sometimes the sunsets begin a little after lunch, or even while it is still happening. Another day spent and another long night ahead, wet leaves picking up the orange lamplight.

The winter solstice, the shortest day, has become something I count down to as the year grows grim and dark. Because this is when the year turns. After December 21 (or thereabouts, sometimes it shifts onto the early hours of the next day), the days will get longer by a few minutes at a time. The afternoons will still be dreary in January and drab in February, but the days are growing. Even when the weather is uncanny and out of season

– weirdly hot in October or suddenly freezing in spring – we can count on the promise of more hours of daylight every day until June 21, when it starts to wind down again.

It's the hope of what's to come that I cling to. It's why I find joy in the winter solstice where others see only misery in the extended gloom, and why I find the summer solstice always lightly painted with melancholy; I see it as the beginning of the creep into darkness and nature's necessary retreat rather than a night of bacchanalia. It feels strange to herald it as the beginning of summer when I know the next three months, while warm, will also be full of ever-shrinking days.

This time-travelling of gardening – to imagine months ahead – offers a balm. It feels like a magic trick, and one that only gets better with knowledge and experience. That, if you know enough about the plants in question, you can stand in a garden and look at it in the depths of winter and see a vision of lush foliage and frothing blossom, opening flower buds and autumn-painted leaves. It's a sober hallucination built from anticipation, science and sure-footed faith, much needed when the basic questions of life feel towering.

I've always been someone whose dreams lie in the near-future, fascinated by the mysteries of what will be and how, and whether I'll remember my earlier trepidation when the events finally come to pass. I spent a childhood vaguely frustrated that, when I tried to make or draw something, what appeared at the end never quite looked like it did in my head, and so much of life thereafter panned out like that. I could never grasp the satisfaction I had been promised.

And yet with gardening that doesn't happen – to me, at least. Even the predictable things, the fattening teardrop buds of a *Pelargonium*, the surety of a green bulb tip pushing up through cold, packed earth, are more wonderful in reality than in imagination. To come home and find a half-inch of growth on something you'd vaguely forgotten about is never not astonishing. Not only cheering but so of its moment, a feeling that resists capture. The lightness of it, the simple pleasure of it, is so innate as to be one of those increasingly rare happenings that only takes place offline, and often alone. A dollop of quiet, personal happiness that is far more difficult to find at work or at home or even in love.

And when they don't work – when the bulbs are blind, the shrubs don't flower, the colours collide in the wrong way or there just isn't enough sunlight for things to grow beyond limp, straggling shoots – the disappointment of it arrives as a challenge, a mystery to unravel. I try to suss it out, establish if the quantities of water or feed were enough, if those plants were put too through much or given too little. I'll chalk it up in my head and try to adapt the practice a whole year later, all the while time-travelling to see better results.

To me, this is a mark of patience, something I still possess so little of. But what patience I do have I have learned through gardening. Part of the reason I got so swiftly hooked on pottering around the balcony is because it quietened my mind like nothing else. Others claim that exercise helps give them mental space, that running or swimming or climbing helps them to make peace with the goals they have not reached. For some, release comes with mindfulness or meditation. As our lives have become busier, we have developed ways to escape the technology and speed that drives them. But I've never known any calm like that which I have gained from gardening. It diffuses my angst, my anxieties, my worries. It stretches time. I follow the tasks that need to be done and take comfort in the fact that some of those results won't appear for weeks – or, indeed, at all. That much of this life finds its own way.

There is, though, some trickery to make things grow faster. People have 'forced' bulbs for generations, putting planted bulbs in cold, dark places in autumn for long enough to make them think winter has arrived and then, as they start to grow, bringing them onto a warm, sunny windowsill to induce a fake spring three months before the real one arrives. The blossom of a hyacinth, paperwhite or amaryllis at Christmas, a riot of flowers inside while frost grips the pavement. But even this pleasing activity requires several weeks of time-biding to be done properly: my forced bulbs rarely bloom on 25 December. There is too much faff at that time of year, the tinsel and the trinkets and the warm glow of the fairy lights. Instead, I welcome their heady scent (for hyacinths and paperwhites are so richly perfumed) in the cold dint of January, when the tree has been taken down and resolutions niggle at our festively softened bodies.

Patience is a kind and gentle thing. It needs nurturing and it will come in its own time. And that January unfolded like the buds of the paperwhites that sat against the condensation of the flat's iron-framed windows. Creased petals pushing against the palest green gauze until, during those long nights, they appeared: starry-shaped and precious and perfectly formed. On a few rare days dawn and dusk bore the stretchmarks of growth, bright streaks of coral leaving pastel clouds and endless vapour trails. Clouds of breath lingered in the air, puffed out between scarf and glasses. But mostly, those first two weeks were heavy and dark: damp, drizzling things that got into the bone. Noon – the brightest hour – passed like a ship in fog, barely noticed.

Plans had been vaguely hatched to sever the contract that still bound Josh and me together, that he would take the flat and I would go elsewhere to start again. I had come around to the idea, but the details took time to iron out. The admin jostled against the heartache for attention; I forced myself to be distant with him, to resist slipping into the patter more familiar between us, the one built from our own language. I still felt guilty for falling in love with someone else, and the weight of it was heavy. This was a happiness I didn't feel I deserved.

And such a happiness. Sunday nights would unfold like bad rom-coms, Billie Holiday on the stereo, smell of toast in the air. His nose would fit in the gap between the bridge of mine and my forehead as we danced, heads touching, smiles indefatigable, sweet-sickened with the cheesy glee of it all but not wanting to stop making those small circles on the living-room floor. Matt was easy about letting me into his life, introducing me to his friends and family and taking me to the places that had helped inform who he was. I was still nervous about letting him see my raw edges.

We did things so differently. A trip to my parents' house was strictly off the cards, but his lived in London and regularly hosted family lunches. Where a trip back to the countryside involved train timetables and routine, we ambled over to his after a kind of notional agreement that we would get there after an impromptu walk across the Common. I was unfamiliar with this land, tended to mostly keep to London's south and east corners. And on that brisk, dry day the space of it felt so wild, somehow so untouched. There was an ochre haze to it all, in spite of the lack of sun. Huge verges of bracken sat gold and rusty, feathered leaves had been rendered crispy. Winter-weary grasses hushed in the light winds. As Matt and I walked its muddy paths, dodging puddles, covering rubber soils in wet, compacted earth, he told me his own history with the Common, which he had grown up near. This was a space of bike rides and Christmas walks, as familiar to him as the footpaths that linked the fields of the village I grew up in were to me. It was here that he came to walk, and then run, after a break-up, clearing his head in the fresh air of open space, letting nature give him the room to think, a necessity at first and then a habit as the pain eased and running became something else entirely, a new challenge, a time to beat, an injection of endorphins.

When we were halfway across, he threw me a playful look. 'Shall we go over the back?' You could, he said, sneak into his parents' back garden by following a cut-through from the footpath. It was a ludicrous idea, the stuff of childhood fantasy. We were dressed up for lunch, not scrambling. But this is where Matt pushed me to, why being with him was so addictive; regardless of how much I wanted to control things, to portion out my feelings and present the right kind of person, he lived by whim.

He defied whatever boundaries I tried to build, showed me the freedom that existed beyond them.

•

For someone who is nearly always restless, January has long been frustrating. Now that I had become near-dependent on that tonic of growth, of taking in the little scraps of nature smuggled in among the city, the depths of darkness and dormancy that start the year felt like a kind of claustrophobia. While my friends embarked on new self-improvement challenges – 'Dry January' and fitness regimes and ambitious resolutions – I couldn't think of much I wanted aside from to feel better. The short, cold days drove us into screen-time hibernation. I felt like it fried my brain.

Living at the flat meant I was reunited with the balcony, and I was keen to tend to it, see what had survived during my absence. Many things die back during December, but the efforts I had put in over the autumn – the bulbs I planted, the cyclamen and pansies in the window boxes – should have lasted, should have been busily preparing to reach their prime.

Still, it was difficult for me to find the daylight hours to get out there. When I did, the maintenance was slim. In larger gardens, January is a time of admin: you scrub pots and clean greenhouses; sharpen tools and prune fruit trees. Such tasks didn't arise in my concrete box. I had no plots to dig over, there was no Christmas pine to mulch let alone an apple tree to trim. There were no leftover pots to scrub, they had been filled with bulbs whose identity I had forgotten. I was grasping for meaningful, physical work; longing for the meditative calm that gardening could bring. But there wasn't much to be done to satisfy my twitching thumbs.

Beyond the balcony, most of nature was in a kind of hibernation. The perennials had died back; many tidier gardeners had cut the dead, dry growth to the ground. The plants that would return were bunkering down, gathering energy while the ground hardened. And, in theory, this time of year is when those in tune with nature take rest, too. But I was yet to learn how to do that. I felt incapable of relaxation, panicky at the sparseness of my diary, vaguely irritated at the flickering candlelight that had, weeks earlier, sparked cosiness. I was fed up of nesting under blankets; a dull ache appeared in my lower back from lying down awake for too long. I was ready for a walk; my legs were desperate for a stretch. Torn between past and future, the present left me somehow frenzied, keen to move at a faster pace, to be productive, to do something when my brain woke me up long before dawn with spiralling, un-clutchable thoughts.

I've never known if this motivation, an inner motor that always seemed to be spinning slightly too fast, is a product of my genes or my generation. I am of fidgety stock: my parents have always created things alongside more responsible jobs that kept us well looked after. My father fettles, thinks, and makes and mends. My mother has always sewed and baked and made and conjured alongside the constant juggle of running a family home and working as a teacher. I grew up around brick dust and projects; I would come down to the kitchen on a Saturday morning to find one of them whipping up a pudding while booking a holiday and hear the faint slam of the garage door as the other got on with lightly frenetic pottering. To-do lists, written in pencil and sprawling with annotations, would be pinned to well-ordered piles of paper that hung off a hook. Known as 'The Clip', at times it felt like this was the family's guiding nucleus, the oracle that kept us all bustling through our various abstract regimes.

But even against that foundation, it feels as if busyness is the constant of my generation. It started to kick in at university, as queues formed outside banks and talk of the credit crunch dominated news reports and a vague pressure crept into our student days, an extension of the extra reading and hobbies that were picked up and written about on university applications. There was always a sense of fierce competition, that one had to be better informed and autodidactic enough to deny another well-read, confident eighteen-year-old one of the handful of spaces to read English at a Russell Group university. We were young when we learned that working was good, and not-working was bad. That jobs weren't a means to buy comfort but a calling, a destiny that we must work endlessly for. That a career must be an extension of our personality, or else we had failed.

University became an increasingly expensive, transactional thing, the courses we took not studies to expand our brain but a necessity to begin that dream-inducing career in a fearsome market shrunk by economic disasters and shopped-around debt. That three-month-long summer after first year wasn't spent lazily drifting around Eastern Europe, as had been advertised, but running errands in the offices of magazines and newspapers between taking shifts in a loathed retail job. It set up a metronome of busyness that would come to define my adulthood – that the given job in hand isn't enough, there must be something else, something to fuel your passion. After all, why bother otherwise? With sea levels rising and the inequality gap widening, if we were not doing something worthwhile even to us, what was the point of doing it at all? We all started to scrabble for so many things beyond employment – interests and hobbies. Even the prospect of relaxing, of doing something fun, became a kind of project to be fulfilled, documented on social media and mined for digital attention.

When I was a teenager, all I wanted to be was a music journalist. I pored over the *NME*, drinking in the words on every page as if they would spell my freedom in some fantasy adulthood, where I would live in London and go to gigs and interview musicians. And I had done that; I had got it. But almost exactly a decade after making that dream a reality, I realised it had transformed into something else: listening to music, finding new bands, the things that had conjured my ambition in the first place had become just other ways of working in my spare time. The books and films and art exhibitions I had found such inspiration in when I was younger had turned into things I had to tick off on a list, to pretend to know about, to harbour informed opinions on. I would look at those in the crowd at shows I was reviewing, both envying and eye-rolling at the fun they were having. I had lost the feeling of knowing what it was to see it, to be there for pleasure. To be a fan with fervid opinions that didn't have to be boiled down and offered up. I became fractious and bereft, infuriated by the fact we apparently had everything on a plate and yet so little to eat.

The need to slow down is something that most people have longed for at one time or another but the term became a brand only a few years after millennials started being born – in 1986, when Italian activist Carlo Petrini became infuriated at a McDonald's opening in Rome and started the slow food movement in protest. Since then, all sorts of slow things have appeared – money, parenting, fashion, travel, gardening, television. Many of them maintain similar principles: a consciousness about one's actions, an appreciation of process and a resistance to mass-produced, unsustainable existence. Slow things are meant to conquer the nemesis of time poverty, the feeling that there are never enough minutes in the day to get everything done.

Among my generation, the slow things became hashtags on Instagram and search terms on Pinterest. To #liveslow would – pictorially at least – involve a coffee with froth nudged into the shape of a heart next to an expensive magazine and photographed from above. Slow cooking was a stew that had sat on the hob for hours, uploaded for digital appreciation. To travel slowly was to take a photograph of something that wasn't a tourist attraction and explain why it made your holiday better. Having grown up with screens that become ever more hungry for our attention, our adulthoods have become ones of scrolling saturation. Our jobs, our homes, how we spent our time all became fodder for online content: what even was it to go on holiday without turning it into an opportunity to Instagram?

And so we started to pursue ways of experiencing life that are no longer dependent on technology. Our adolescences were spent in the shiny, gaudy opulence of the Noughties: music videos swagged with 'bling', footballers wives leading the charge with wildly expensive handbags and the celebrities we idolised wearing velour

But that changed. Within five years, people – my friends, friends of friends, colleagues and strangers on the internet, nearly all of whom were my age – came to talk to me about plants. They had the same concerns I had and the same desires too: to bring greenery into their lives, learn how to look after it and be reassured that it wasn't dying. I would often prescribe patience and emphasise that there was nothing wrong with their no-longer-brand-new plant at all. Alternatively, they had often overwatered it. Even with the limited time and experience I have clocked up, I have gained some of that gardening philosophy that the plants will revive if given the right light, water and opportunity. That a bit of a brown leaf tip does not mean imminent death but that the central heating has been on too high. Usually, the best thing is to wait and see and take delight in the process of watching it all happen at its own deliciously independent pace.

The shift towards our generation finding gardening – whether tropical house plants kept indoors or an enthusiasm for growing our own food, even in window boxes – for ourselves happened alongside a wider trend for plants in fashion and interiors. When Scandinavian womenswear brand & Other Stories opened a long-awaited branch in Regent Street, the racks were separated by *Crassulas* and *Ficus lyrata*. The plants cultivated a desirable, aspirational air: one of space and time and an environment where life could thrive, even under the punishing lights of a retail space. East London design brand House of Hackney set the tone for this some years earlier. When it launched in 2011, its luscious, foliage-based Palmeral print immediately became the brand's unofficial trademark and allure. House of Hackney based its identity on Loddiges, a Victorian nursery of exotic plants that in the 1820s boasted the largest hothouse in the world. It was integral to introducing exotic plants – including orchids and palms – to Europe. While all that remains of Loddiges are two palm trees outside Hackney Town Hall, where the nursery stood two centuries ago, a string of designer plant shops have emerged in the neighbourhood, once again bringing tropical plants to the plant-hungry millennials who live in the borough.

Our desire for steadiness is not new; we are merely the latest generation to nurture it. I can't walk past Liberty, the black-and-white mock-Tudor department store that crowns the bottom of Argyll Street, without looking at the ferns that adorn its doorways. They are relatively new additions; they cropped up a few years ago, jungling up the pavements trodden by sandwich-carrying office workers and flustered tourists, flanking the endless queue of black cabs waiting at the zebra crossing with a surreal green. For me, the presence of these ferns pinpointed the moment that plants returned to the mainstream, when they shifted from frumpy to fashionable, inconceivable to cool, and transformed into as much of a symbol of desirable living as the £90 candles stocked within Liberty's doors. It also makes me think of how we are stirring up another Arts and Crafts movement – a second wave, a century or so later, driven by a similar desire to slow down and turn our backs on the technology that has left us as one of the most overworked and vocally anxious generations yet.

Along with Heal's (another London shop that started to sell terrariums and other plant matter in the mid 2010s), Liberty was a Victorian department store that supported the work of artisans in the Arts and Crafts movement, a sprawling group of collectives that spanned the practices of architecture, design, gardening, art and crafts but were united in their ambition to live more simply and find inspiration in nature. After the frills and the fumes of the industrial revolution, the Arts and Crafts movement looked to find a more fulfilling way of life by returning to handmade, purposeful objects and the gradual pace of the natural world – essentially, a time before industrialisation. They had other radical ideas: that men and women could be equally vital in creating and consuming thoughtfully designed objects and that the lines between practices could be happily blurred, so that a painter could be a gardener and have an input on the architecture of a house (looking at you, Gertrude Jekyll) and that manual work, or using your hands and body to achieve things, could be satisfying.

Such Arts and Crafts rebellion is built into Liberty's timber-framed skeleton (itself taken from the skeletons of ships). In 1884 Edward William Godwin launched the shop's Costume Department. A bit like the designer concept stores of today, this was an extension of the ideals with which Arthur Lasenby Liberty established his eponymous business: to prompt change by pushing boundaries. It wasn't merely a place to sell clothes, but to educate customers about new ways of doing things while providing 'the most beautiful types of modern dresses' for 'amateurs, artists and the stage'. Godwin wanted away with Victoriana's restrictive corsets and dull suits, instead bringing in loose, brightly coloured designs in rich patterns inspired by nature. William Morris, Arts and Crafts' lasting poster boy and one of its most influential members, collaborated with Godwin and Liberty, who stocked his designs. Morris's work combined petal and leaf in bold and beautiful colours and was turned into fabrics and papers laboriously by hand. We still hang machine-made versions of his designs in our homes nearly 150 years later. He championed the inherent appeal of nature, entertainingly believing carpet bedding – the habit of planting flowers in a gaudy, compacted mass to create artificial shapes and patterns when seen from above – to be 'an aberration of the human mind'. To make a garden look like a natural landscape had long been sniffed at, but this rash of new designers wanted to change that. Before Piet Oudolf and Henk Gerritsen filled gardens with plants that looked bewitching in death as well as in life, members of the Arts and Crafts movement deployed perennials in the land that surrounded their innovatively archaic homes to create the pretty ramblings of gardens that embraced nature. Foliage was celebrated over the flashiness of flowers. In 1867, a plucky twenty-nine-year-old gardener named William Robinson started a small revolution by filling Battersea Park with tree ferns. He suggested the softness of pampas grass, the structure of yuccas and the delicate spears of bamboo for domestic gardens that had previously been awash with far fussier flora.

So many ideas from these frustrated and idealistic Victorians feel familiar. We may have fewer gardens to plant but we also hark back to simpler, more optimistic times. Mid-century design has become fashionable again; we lust over G-Plan sideboards in the face of Ikea's convenience. We usher rattling faux-marble drinks trolleys and low, utilitarian sofas into our homes. These are the trappings of the 1950s and '60s, decades of promise and Space Race optimism. Crucially, a time before the internet existed.

Craft has proliferated. The word has become synonymous with care and attention, something that became so abstract during the money-lusting decade in which we came of age. 'Craft ale' has exploded in popularity, ceramics and woodworking have returned to vogue after years in the interiors fashion hinterland. Much as the proponents of Arts and Crafts encouraged beginners to get involved in making things, with organisations such as the Home Arts and Industries Association existing for exactly that purpose, so potteries and floristry studios have welcomed professionals from other walks of life to take part-time lessons as millennials pursue creative endeavours alongside their day jobs to find pleasure away from the daily grind. To produce something with their hands as a relief from jobs where hours and meetings and emails have little to show for them but overdraft relief.

This, in turn, takes us to the 'slashies' and 'side hustle' phenomenon, or the increasing acceptability of people pursuing passion projects on the side of their nine-to-five career job, challenging the traditional notion that a person should have one type of career and stick to it for life. Instead, fed up with the confines of the 'dream job' we spent the first decade of our careers trying to achieve, my generation are trying other hats on. Sometimes they stick. Often, the extra jobs we take on allow us to make or do things away from a screen: create a pop-up restaurant or conjure something physical out of nothing but time and creativity. Get out and dig.

Passions that become professions live beyond the contracted forty hours a week. The DJ / Model / Influencer is the oft-mocked millennial stereotype, but there are others who are getting on with two or more pursuits without many of their traditional colleagues realising – a PR / charity worker, for instance, or an events organiser / potter. Here lies a contemporary re-imagining of the boundary-blurring that the Arts and Crafts movement adopted when they applied their creativity across traditional borders.

The slight snag is that while we pursue different, more creative, more, well, *artisanal* pursuits, most of us have to keep up the day job. The Arts and Crafts lot tended to set up short-lived communes in the Cotswolds – now, if we're very lucky, we might stay at Soho Farmhouse. Our generation are finding new places to live beyond London, which is one of the reasons behind Margate's mini regeneration. The notion of 'slowing down' continues to prove elusive and, to a certain extent, unwanted. We still want a cab at the click of a button and our online shopping delivered within the week. Those side-hustling crafters will nonetheless upload their efforts to Instagram (#makersgonnamake). But then, it didn't all work out for Morris & Co either – their handmade products were so expensive they were caught between serving only the luxury market or struggling to exist at all.

For all my impatience, I did manage to find small ways of slowing down amid the stultification and cabin fever of that cheerless January. Its middle weekend started with blue, fluffy-clouded skies for the first time in weeks. I wrapped myself up in layers of beloved, reliable and thoroughly unflattering thermals and fleeces, felt the heavy soles of my scruffy walking boots pull gently at my calves and jumped on the bike to cycle down the hill and over the rise of Brockwell Park to the community gardens. It had started to rain by the time I'd reached the greenhouses, and there were few other volunteers around, which meant I was thrust two cardboard boxes of tulip bulbs and told by the head gardener to get planting, as 'they're better going in late than not at all'. It was so reassuring – you're meant to plant tulips before the first frosts harden the ground; I would always try to get them in by Bonfire Night. But here was a pleasing battle between logic and gardening law: better to put something in with a bit of hope than waste it altogether.

Some of the bulbs were already sprouting, keen to get on with a flower that was yet to be rooted in the ground. I set to, using a fork to test the soil for hidden lumps of concrete and then digging shallow troughs under trees and along borders. After trying to squeeze half a dozen bulbs at a time into one container, the freedom of scattering a couple of hundred about felt ridiculously opulent, like throwing confetti into a duck pond. I nudged the cold earth back over the bulbs, which I'd placed with flat bottoms and small roots against the earth and pointy bits up, tapped down and drenched them with the hose. My movements were compromised by the two pairs of gloves – one woollen, another borrowed and specifically for gardening – and the condensating drip of my nose. Signs of a person too used to being indoors; the head gardener's hands were bare and fine, plonked into the icy water butt to rinse off at tea-time.

I've always seen bulbs as something alchemical. Brown and crispy onion-shaped things, their tracing-paper skins can inflame hands and yet be so delicate as to be left discarded in the bottom of the (ideally) paper bags they arrive in. But no bother. They do not need this most translucent of coverings, for they hold drainage. I have brought paperwhites to bloom that have wrapped their roots around nothing but gravel, because the bulbs, like eggs, are neat packages of self-sufficient energy. All that life will emerge from them with mere sunlight alone, even in the darkest, coldest days. Let them die back properly – leaving the green leaves to yellow before cutting back – and they will return the next year, maybe not as vibrant, maybe softer and more graceful, but maybe even better for it. And nearly always when you have forgotten that they are there. It doesn't take many seasons of gardening to be surprised and mildly irritated by the presence of the one rebellious, wrong-coloured tulip that you thought you had uprooted the previous summer. Really, the main thing bulbs need is time. They will sort themselves out, forcing shoots around the other things lurking in the soil – even ones that go in upside down can manage, making excursions around themselves to complete that survival homing mission of bursting into the clean, cold air above. Their reward will come if granted several weeks – months in the case of the grander ones – of slow, understanding patience.

With time, too, I had become better at looking for signs of spring in January: a more grounded kind of natural time-travelling. Knots of small and shiny, chestnut-coloured leaves sprouted out of the stems of hydrangea bushes, their flowers reduced to nodding heads of dry petals. Rashes of pastel pink *Viburnum* blossom injected leafless streets with a sweet-swelling blush. Hard, spiky pom-poms dangled from the plane trees that lined London's streets. *Mahonias* interrupted swathes of sullen green with spires of acid-yellow bells and careworn verges would pop with snowdrops – first one or two, then a small drift, until the whole stretch had been painted with tiny white polka-dots. Few things were in bloom, but plenty were getting there: the frost-caked soil of the balcony tubs was increasingly interrupted with the green shoots, signs that bulbs were reaching the soil. The days grew longer, sundowns warmed into glittering, golden things – even if they happened long before dinner.

And in those dark hours gardeners stay indoors. For once the maintenance and the tidying is done, January offers the space for reflection and planning. In more organised households, seed catalogues arrive and are hungrily pored over, like people looking wistfully through pictures of their summer holiday in the depths of winter. Colour schemes are thrashed out, varieties ripped from pages and rearranged on the kitchen table with a glass of wine. Those with drawers of seeds sort them and see what might be sown. From the warmth of the inside, whole gardens are dreamt up, mistakes are remembered and remedied, new plans made and calculated. While others run their resolutions into sweaty trainers and resist the pub, gardeners vow to do certain things better this year, for the chance to remedy last year's errors only comes once every twelve months.

I had not acquired many seed catalogues at that point. One needs to get on mailing lists – whether by ordering from specialist nurseries (something to graduate into, after the stages of solely picking up plants in supermarkets and impulse-buying from local nurseries) or being organised enough to sign up, and I hadn't done either. But I made do with metaphysically growing things, reading from grown-up picture books about plant-filled homes and feeling the prospect that things would thrive and be green again smother my soul like a salve. I made lists, plotted and schemed, time-travelled my way through a spring and summer on the balcony without even knowing if I'd be there to tend to it. It was reassuring work that helped me make peace with the fact that I was leaving the flat – and the balcony – in a way that I had often entertained as a worst-possible outcome: moving my things out, alone, leaving only memories and severance behind. Because now that was on the horizon I realised that it was a cessation that was necessary in spite of all its upset. I had been drifting for so long, caught in a miasma of guilt, loss and uncertainty, that even this thud of closure held something of a relief. Like bulbs beneath the earth, or perennials in dormancy, the stasis I had been in was nudging towards a sense of conclusion – even if the notion of how that would logistically unfold, and what would happen after, remained blurry.

Josh would get the flat in the long run, but I was granted it for the few months it would take me to find somewhere more permanent. It gave me the promise of a couple of final seasons on the balcony. And, knowing my months were numbered – and unsure of what growing space I would have next – I decided to have one final hurrah in that concrete box in the sky. I set to growing things, started sowing seeds. I wanted to push life into that space, no matter how temporary, to see what would grow. After months of encountering nature as a kind of tourist – drinking in woodland; understanding the need for dormancy; finding breath in green lungs; unravelling plants' tiny mysteries and seeking the protection of walls and glasshouses – I wanted to participate in it more fully, to create its existence and thrive in its making.

There were still plenty of unknowns: this was far from instant gratification, a needed hit of indifferent greenery to cut through the difficult days. But I wanted to participate in the process of caring for, nurturing and feeding this space, even if I didn't know if I'd be around to see it blossom and fruit. And that was all right, because with every plan I made and every plug I planted, I would imagine what it would look like in the months to come, beyond the realms of my unpredictable reality. Minutes of peace that existed only for me, in a world where everything else was on show.

FEBRUARY

Excitement lay in tugging open the bike shed door, hearing the familiar scrape of its damp wood on the tarmac. I'd given myself an extra few minutes to pump up the bike tyres, getting them hard and strong and ready to rattle down the hill, breath steaming up my glasses.

Cycling had been a source of discovery and distillation – of thought and time – for several years now, but the bike had proved a tricky thing to store in those temporary homes. As I became more adept at whittling down the things I needed in an attempt to establish what I actually wanted, the bike – a scruffy, rusting, green road model with suicide brakes – stayed safely in the shed on the estate, tyres deflating into soft rubber. Learning new commutes in the depths of winter proved just another insurmountable thing in those darkening weeks of difference.

But to be back on the bike granted fleeting, familiar escape. The distances I travelled were short; a few miles to work along the major arteries of Camberwell and Vauxhall Bridge, down the hill to the community gardens, across the park to Brixton, where I would chain it up on the rails outside Matt's flat. My earlier years in London were divided into chunks of cycling time: an hour to work, another one back, up the eastern curve of the city to visit friends in Hackney and into its beating heart for the office.

While I no longer needed to span such stretches (I could now afford the train and was reluctant to cover my clothes in sweat and street grime), the mechanical pace of foot on pedal pushing wheel on asphalt still gave me clarity and thinking space I only otherwise managed to find when gardening. They shared the similar need to concentrate on the captivating task in hand, of combining simple task with physical function. Among the exhaust fumes and the red lights my head would clear and be filled, instead, with words and musings, new ideas and daydreams. I'd arrive at my destination some twenty minutes later with a beating heart, the thud of blood a more present rhythm beneath my skin. It had been that way since I was twenty-two, when I was scared and emboldened in equal measure, existing on appetite and air. A Cupidian arrow over the fearsome walls that stood between me and London. It let me into its secrets, gave me a sense of collaboration and ownership, allowed me to read the city by its seasonal changes – how the wind moved the rain and the incremental changes of the sun's rising. For a city that is so much the preserve of the rich, London offered plenty to the skint, as long as they were inquisitive and on two wheels. Here lay a kind of freedom I was fully in control of.

When I didn't have to be anywhere at a certain time, I would let curiosity guide me, taking turns down unknown streets and ambling slowly along, gazing at the unfamiliar houses and shopfronts, safe in the knowledge that I'd find my way back. After the end of my early twenties I'd grown out of the habit of exploring; my diary became a hectic thing stuffed full of drinks and meetings and catch-ups that would be cancelled, rearranged, then cancelled again in a flurry of apologies. Overwhelmed and short on time, we'd spend the days quietly hoping those in our calendars might renege on plans first, so we weren't the ones to bail. It was an uneasy balance, being too exhausted and anxious to socialise yet so caught up in the need to keep up with friends and appearances that we padded out our lives. Increasingly, I felt the need to get home, where I would tidy and cook and wait for Josh, the evenings stretching out ahead of us, to be consumed by television or some other screen.

These pressures had moulded into other places, now. There was nobody to go home to; my plans had become shapeshifting things. There was more need to be curious, more space for idle wondering. I wanted to discover London again and purely for the sake of it. One clear afternoon in February I took a turn off my regular commute and into Bonnington Square, a clutch of intertwining Victorian terraces smuggled behind Vauxhall Station. It makes for easy discovery by bike; the streets are barely bothered by cars and there are humble wonders to be absorbed. The houses around Bonnington Square are an urban gardening paradise, a vision of what might happen if we deliberately greened our pavements and porches with time and effort and wild abandon of what the neighbours might think.

I was entranced; where one would usually find pavement or parked car there were spacious, jungly beds. Huge tree ferns leaned over kerbstones, underplanted with a gaggle of green shoots where daffodils would later flower. The fronts of the Victorian houses – handsome wooden windows and bricked arches – hid behind little arboreta: tropical trees and rhododendrons side by side with woolly *Cupresses*. A catalpa lorded over a corner of houses, a carpet of its shed seed pods covering the pavement like crispy aubergine skins. Fan palms serrated the blue sky above, sheltering a *Fatsia japonica* that still bore the atomic heads of bloom. Huge tufts of grass exploded from gaps in the paving stones, made pale and brittle by the weather but still beautiful in their sheer presence.

Beyond these generous beds lay others made of pot and crock, of box and bin and saucepan and watering can. The sword-like dark green leaves of a black mondo grass unfazed by the winter, probably because this place had developed its own microclimate from brick and bark. A little fig tree bore leaf in a pale green box that gave a long-ignored warning: 'PROPERTY OF NHS LOGISTICS'. The leaves of an aspidistra erupted from a steel dustbin; the pink-and-green foliage of an adventurous *Tradescantia* poured out of a plastic window box and mounds of *Muehlenbeckia* – that truculent Antipodean maidenhair vine – masked any container it presumably inhabited. Bulb shoots ripped forth out of a froth of green dwarf *Oxalis triangularis*, which was colonising all nearby cracks in the pavement. Every now and then, the tropical hardiness of it all would be interrupted by a pastoral oddity: frilly pink petal tips emerging obscenely from a fattening camellia bud; a crop of yellow primroses neatly filling a tub. One concrete pot, smothered in anonymous bulb shoots, bore a small label: 'Chantal & James, wedding nectarine, 1993'.

The beauty of these gardens – and to me they were gardens, even if there was no lawn, no beds, no planting plan or uncontained soil – lay in the fact they existed at all. To look at them, even with an untrained eye, was to realise that these little forests, these layers of life process, had been planted with intention beyond certainty. Because these trees were both an attempt and an optimism that had a longer life than the houses they surrounded; had it not been for a few hundred disparate people, Bonnington Square wouldn't still be standing. Like the crane-filled land that surrounded it, it would have been filled with shining flats waiting for overseas buyers and offices crammed with people who would later commute home beyond the city. But the houses were saved through a mixture of goodwill and survival, and gardening was integral to maintaining their inhabitants' fragile claim on those streets.

In the late Seventies, the hundred houses surrounding Bonnington Square were desolate and bleak. The windows had been bricked up, the doors boarded over. Inside, floorboards rotted above rusting pipes, remnants of wallpaper fell to the floor revealing the plaster and lives lived there by those who had abandoned the properties. Young people looking for somewhere to live and something to save would cycle around London's increasingly divided and decrepit streets, trying to find empty houses. Eventually, just like I had, they found Bonnington Square. One newcomer described it as being like a fortress.

People broke in – through smashed windows at the back or with crowbars through the front. For some, this was a hasty endeavour: the more quickly they could bust in and change the locks, the faster they would have squatters' rights over the place.

But for others, opening their own front door was a symbolic act, no matter the time it took. One woman unsealed hers with a crowbar and the commitment of her whole body. 'Each nail was one step closer to having my home,' she later recalled. This was a practical action imbued with meaning and significance, a need to roost, a desire for a kind of permanence at a time when so much was defined by instability and change.

The community grew quickly, attracting people from all over the world as a means of quite literally making a home in London. Pipes were plumbed, wiring was done, window frames and furniture and even chandeliers were scavenged from skips and used for nest-feathering. And the plants came, too. To look at photographs and footage from that time is to see spider plants and philodendrons and Kentia palms and *Pelargoniums* sitting on window sills and in the corners of makeshift kitchens. Unbricking a window meant opening up the space to the park beyond. Roofs were turned into gardens, with troughs of tulips dancing against the sky above. Gaps in the pavement were filled with greenery. Before people had even properly sorted their plumbing or electrics, they had made space for greenery.

When garden designer Dan Pearson arrived on the square on the cusp of the Nineties, his roof garden overlooked a patch of wasteland where seven houses had stood before they were bombed in the Blitz. A lone walnut tree, planted by some of the square's first squatters in 1983, was surrounded by bindweed and buddleja and the remnants of play area equipment optimistically installed by the council in the Seventies. A chain-link fence separated it from the terraces beyond.

There was a sense of that kind of in-between here, the sort that exists in verdurous places trapped by cities. Sirens screamed past – the garden was flanked by a heaving artery between Camberwell, Oval and Vauxhall, which I cycled along most days – but the brick walls and stone underfoot boasted the green moss of somewhere with far purer air and stiller surrounds. Purple croci carpeted an area where a bench lay vacant, waiting for warmer weather. The stone paths petered out, tempting adventure into other nooks cut off by hedge and fence. There was nobody there but me; most people are working on weekday afternoons. I stood there a while, feeling both welcome and intruder, grateful for this pocket of green and bemused that I'd never visited before. Even before I knew its history I was so conscious that this was a space that had been created out of a playful, primal urge, to make something that was peaceful and beautiful for its own sake. This wasn't a public garden that was an add-on to justify another high-rise of luxury flats, but to make good from rubble, to give a green lung to a blossoming community. A space for quiet and nature and learning in the middle of the city. Another kind of controlled freedom against London's towering walls.

It was one of those lingering midwinter days where, just for a few minutes at a time, whispers of spring drift through the air. The sun may be low by mid afternoon, but it takes its time to set, making crisp skies gentle and deep as the cold sets into knuckles. Raindrops scattered from cloudless skies, the street lights seemed on slightly too early. A daffodil in bloom would still seem odd, but a snowdrop was beginning to appear, a relic of a colder, darker time. Green shoots of eager bulbs punctured the soil; bare branches withstood strenuous winds. After weeks of January's stultification, a new month had arrived with tremulous, unpredictable energy. It was wise to revel in the weather that arrived when it did, for it would change in a heartbeat.

My renewed determination to garden the balcony in what little time I had left there had made February fruitful. The white cyclamen planted in October were still holding on, bright in the boxes amid a brittle blanket of ivy, but to them I added Columbia Road Market treasures to fill in the gaps: ranunculus and some dark and regal hyacinths on the cusp of blooming, flowerheads tightly packed within the nape of the leaves. I'd planted hyacinth bulbs in other troughs – white ones, set to pop up through a mist of silver artemisia – but those were only just gingerly pushing the smooth green domes of shoots through the surface of the soil. It would be weeks until they flowered, and my patience had run thin.

There's something blowsy about a hyacinth. Perhaps it lies in its famous namesake, Hyacinth Bucket, the insufferable star of class-skewering Nineties sitcom *Keeping Up Appearances*, but there's also the scent. An intoxicating, room-filling thing that, like Bucket, is an acquired taste. Some describe it as sweet, but it offers a deeper, more penetrating hit than mere sugar, the kind of dizzying smell that verges on animalistic. There's a potency and presence to it, stronger than dreaminess and nearer desire. Like daphne, which also blooms over winter, you can catch the smell of a hyacinth on a gust of cold wind during even the bleakest and drizzliest of gloomy days.

Josh and I were given white hyacinths as a housewarming gift when we moved into the flat exactly three years before. They're bulbs people force for Christmas, but like other bulbs people force for Christmas their appeal, if anything, increases as the parties die down and the solemnity of winter rumbles on. I've always thought them oddly flamboyant for a plant so well-suited to barren winter landscapes, as if they were the horticultural equivalent of a well-dressed urbanite who got lost on the way home from a party in the country and never quite made it back to town.

And I perched these ones on the kitchen windowsill of our new home, where they stood against the condensation of the windows during those dark evenings.

The flat was hot, then – the heating constantly on, thrown in to the cost of the service charge and mandated by an antiquated building management – and the little starfished flowers that make up hyacinths' grabbable stems, once so full and jaunty, soon crisped and turned brown. I didn't know that their decline was part of the process, an inevitability of their impermanence just like any other blooming plant, and fretted over it, conscientiously watering the small pot of soil they sat in and hoping for a revival. Eventually, Mum told me that this was normal, that they had done flowering, and I should put them outside – they might come back another year. She would know – my mother can't bear the smell of hyacinths and any she is given swiftly end up in a dedicated bed beyond her kitchen window. It becomes a riot, a glorious hodgepodge of mauve and pink and white, all forced bulbs that were intended to be seasonal frivolities given the chance to go wild in the great outdoors, to fend for themselves against squirrel and frost.

I didn't put the faded bulbs in the window boxes; it would be some months before my adventures in balcony gardening began.

Instead, I put the little ornamental basket they had arrived in in a corner of the balcony, whereupon it was gradually covered by fruit crate and plant pot and soil and experiment, brown leaves sheltered and kept dry from the elements by negligence alone. A year later, while the days were still short and crisp, I discovered it shooting. I think I'd even forgotten what had grown there; maybe I didn't know the name of it in the first place. But I brought it out into the light and it bloomed again: white flowers, on a spindlier stem and of less vigour, but still enough to coax that near-obnoxious scent into the air whenever we opened the windows to let out the stuffiness. A small dose of resilience.

•

Anna and Heather came round most weeks. The lodger was increasingly out, off with a new girlfriend or out at a bar, and sometimes in but behind a bedroom door that muffled the sound of a television. The three of us would sink into a gorgeous familiarity, of rounds of tea, all squished on the same little Ikea sofa, feet tucked under a synthetic, staticky blanket. Nestled in the pure ease of one another. I knew this was rare and lovely; it was something we had grown together that bucked the fashions of how so many people our age socialised, of how we socialised with so many of our friends who weren't each other. We didn't rely on cocktails or fancy bars; to take a selfie would be unthinkable. There was an unspoken law that we didn't talk about work; our jobs and careers had no place in this girlish, near-adolescent space. One Friday night, after they had gone home, I stayed up alone and drank-in the space where they had been, listening to an old song we played at university. I was so conscious of the imminent loss of this, of how things were about to change. Anna was engaged, would be married within a couple months. Heather looked to the future, felt a tug of stability arrive with her thirties. I knew we'd be growing up soon; I was scared of it.

That fear didn't come out willingly but in a hot, frustrated rage of not-knowing, of feeling the need to be too many things at once. Matt and I were trapped in a tussle for form and control. Each shuffle we made towards comfort or domesticity flashed at me like a warning sign. As we spent more time on the sofa and less, say, grabbing sushi at last orders, I became convinced that such comforts would smother us the way they did when I was with Josh. Matt had been such a welcome blast in my life; it was as if I'd become dependent on the energy he had offered and would suffer crushing withdrawals when it seemed like it was settling into something more recognisable, more normal. He'd always welcome me around his flat but I was vexed at how easy he found it to, say, do the dishes or hang out his laundry while I was there – while I was ripe with anxiety, taking every minute he wasn't spending directly with me as a sign of disinterest. I felt abandoned, would panic that he was getting bored with me, would toss me away like I had been before. I'd lose my temper and struggle to explain why, so would slam doors instead. He'd be mystified, but I couldn't bring myself to explain, tell him that I was worried he had got it wrong, that I wasn't the kind of person he wanted me to be.

It was easier to seek escape. I lay restlessly in Matt's bed, upset building as I listened to the rhythm of plastic keys of his typing softly through the wall and trying to figure out why I felt so rejected when I couldn't even work out what I wanted. One morning I took my bad temper out into the cool grubby air of Brixton and pounded the pavement until I reached Brockwell Park, my rubber-soled walking boots feeling odd within the sea of late-morning commuters. I was destined for the community greenhouses. In those red-bricked walls I had found a sanctuary, a hidden place of industry and calm where little mattered apart from the job in hand. There was process here, physical and seasonal progress that combined logic, instruction, science and luck with time in the hope that something beautiful and orderly could be made from the dirt and the havoc of nature. I was one of the very smallest parts of it, hardly a regular volunteer, nor one with much knowledge. And I liked that; I liked being unintegral, a tiny contribution to a plot few knew about or understood, giving my time and energies to the land in an attempt to make it better.

This was the time of year when gardening is more labour than nurture, especially in a large space where new projects are going on all the time. That morning, the two of us were given spades and wheelbarrows and tasked with levelling a patch of lumpen earth so an outdoor play kitchen could be built there for children who visited the gardens. Not a glamorous activity, but satisfying in its simplicity and purpose. Like much of London, the community gardens have clay-heavy soil. The overnight frost had helped us – when temperatures rise above freezing, the moisture in the soil thaws, causing the clay to soften and clod together, putting up a gluey resistance to the blade that tries to remove it. But the middle of the park. Junk and nature began a silent battle; as one piled up through human effort the other tried to claim the land back. A whole acre of wild space left just to be.

But people began to see the potential beneath the mounting detritus. According to this oral history, squatters were the first to reclaim the space. Like those in Bonnington Square, they started to make good, removing the rubbish and tending to the land. By the Nineties a group of guerilla gardeners known as Green Adventure had fostered the space with the ambition of returning it back to productive land. They planted fruit trees that became an orchard and hatched plans to transform the whole plot into somewhere that could nourish the people who lived nearby.

As time went on, this rebellion force for good started to legitimise. As Bonnington Square's residents also realised, after a while, permanence and protection can only happen with the cooperation of the authorities. The beginnings of a charity that would become Brockwell Park Community Greenhouses emerged. By September 1997 business plans were drawn up for a Brockwell Park Community Environmental Centre, with costs and justifications and timelines and a Green Adventure mission statement that, two decades on, remains familiar:

> To enable inner-city people, in particular those of us who are socially and economically disadvantaged, to create and take part in practical community projects which promote sustainable development thereby empowering us to improve the quality of life for ourselves, our community and future generations.

The hand-drawn plans, too, show the gardens that I had become fond of, with compost heaps and cold frames, a pond and the dye garden. The greenhouses, I realised, had been given nicknames – Esther for the larger one, and Sunshine for the little. In these pages lay the ambitions that I had seen realised: the restoration of the crumbling Victorian walls, the transformation of a 'rubbish tip' into the cosy room where volunteers gathered for cups of tea and cake, the establishment of a native plant-smothered woodland area that trapped and dappled light.

It hadn't all been easy. The early Green Adventurers I could get hold of spoke of far less utopian – or simply more banal – circumstances than those who must have dreamed up this vibrant, well-loved community gardens from what had been a tip surrounded by a beautiful Victorian wall. Some volunteers joined in with the rubbish clearance that was vital right at the beginning of the project, but life with small children got in the way. Small intrigues unfolded, of dramas and jostling and even more scurrilous rumours, drifting and lost to the flotsam of memory and narrative. Others speak of 'small periods of chaos', of lost keys and long meetings about finance. But still, it happened, we reap the fruits of it today, my body working soil that had been sifted through, rescued and fought for, making the paths for new generations to play in nature, just as the original Green Adventure vision had intended.

A few hours later, I stood undressed in the kitchen, having uncovered a bruise on my thigh the size of a satsuma, a remnant from my clumsy handling of space, fork and barrow. Having levelled the patch, I cycled home up the hill and dragged my newly weary body into the flat, where I stood in the hallway and took off the layers covered in mud, sand, soil and sweat, bundling them immediately into the washing machine, letting the heat sift through my prickled skin and into my warming muscles. On the side lay half of an ornate green tile that had been rounded by time and soil before I pocketed it, a treasure unearthed while digging and saved from the skip. Bottle green with pretty raised leaves in each corner, it looked like something discarded from a Victorian fireplace, still holding a hint of its original glaze once it had been washed off under the tap. I ran my fingers over it, wondered about how it had wound up in my kitchen, in my hand, who had discarded it and when, who had laid it, proudly, and where. My earlier tetchiness had shifted, blasted out by the chill and the repetition of movement and from the things I had learned. My body ached but my mind keenly ticked, no longer with agitation or confusion but gentle clarity.

Digging through that soil, and with it, its history, proved what could come from broken land. It was important to acknowledge that improving things that had been left to grow over and abandoned was rarely easy. There were no rules in place: in order for Bonnington Square to become beautiful, it had to be reclaimed using courage and the cusp of the law. Brockwell Park's dump was only transformed into the community gardens by people sneaking in over the decaying walls and clearing it up illicitly. The order of things took place with guts, conviction and determination; there were squabbles about form and process. It was only with time, collaboration and sheer hard work that something good and proper could be established. The trifles and fall-outs and meetings and lost keys folded into a history that lay invisibly among the plants that grew as an outcome.

Matt was due to leave for India for a month, a bit of time to gain some sense of adventure after several years' hard work without much in the way of a holiday. The night before he left was feverish with anticipation; he'd not been on such a big trip. Better suited to dreaming and drifting, he was left flustered by the packing and preparation, the newness of it all. I'd been roped in to help but I lost patience with him, felt we should be spending what time we had together before he went doing something more meaningful. I vented at him, stormed off and ran the bath. A little while later, he opened the door and stepped into the water. I was still cross, infuriated that he'd left it all to the last minute, saw his reliance on me not as a need for support or love but as something selfish. As he invaded the space I'd tried to make a sanctuary, I pushed him away. But this is hard to do in a bath, and slowly the words unfurled, more thoughtfully than they ever had before. I was scared at how comfortable I was becoming to him, convinced that eventually he would grow bored with me, once the gloss had worn off. I told him I thought we might want different things, that I was done with becoming so familiar that I was rendered invisible – that I had better things to do. And he listened, said it wasn't how he saw it, said that I was just becoming part of his life, and he liked it. Gradually we inched closer, limbs getting tangled up as we vowed more patience and understanding, to make things better.

If I was to make peace with the new, softer space that Matt and I were occupying, I had to uncover my history and respectfully let it be, just as I had saved chunks of other, older lives out things that hold the most ambition. A series of steps that would mean my pulling up of couch grass in the community gardens' clay soil those wintry weeks before would allow wildflower seeds to be laid in those to come. For the Blitzing of London's soil and the abandonment of the High Line to give countryside plants a space in the city. For the sheer need of somewhere to live leading to the urban jungle of Bonnington Square several decades later. The old doesn't need to always be done over, but appreciated and nourished. If we allow what has been before, it can become an opportunity for things to linger long enough to grow.

Guerilla gardening started that way. People have been doing it for centuries, although the first official account of it was in 1649, when a hungry textile merchant named Gerrard Winstanley was so infuriated by the illogic of not being able to grow things on the common land that he rallied a group of men and women known as the Diggers. He wrote a pamphlet decrying the fact that 'land which would have been fruitful with corn, hath brought forth nothing but heath'. Within a week he and the Diggers were clearing whole swathes of St George's Hill in Surrey to make room to grow parsnips, carrots, beans and barley.

At the same time, but in New York, every house had a garden and pasture for livestock. The city was nascent, and people could grow things to eat. By 1973, though, Manhattan and Brooklyn had become a drug-addled hub of criminality and decaying buildings. And Liz Christy, an oil painter with cheekbones like palette knives and a Columbia education, was beginning to realise that tomato seeds from the piles of rubbish mounting in where she lived in the Lower East Side of Manhattan were turning into plants, and so other things should be able to grow too. It was Christy who invented the term 'guerilla gardening'. She, with a handful of friends, scattered handfuls of seeds in the vacant lots near her home and planted in the ground beneath the trees. They armed themselves with 'seed bombs', natural grenades made of compost, seed and water, thrown over the fences and into the deserted plots to encourage flowers among the rubble.

Christy's neighbourhood junkyard turned into the Bowery-Houston Community Farm and Garden – New York City's first. Before anything could be properly grown, she and her group of Green Guerillas (Radical Rhizomes had also been a consideration, which I feel wouldn't have been as catchy) spent a year clearing junk out of the plot before soil, manure from the local police station's horses and donations from nurseries started to green up the place.

Christy died tragically young: she was just thirty-nine when, in 1985, she was killed by cancer. By that point, the seed bombs she had grown had exploded into an urban gardening revolution. Christy spread the word of community gardens through her *Grow Your Own* radio programme and became the first director of the city council's Open Space Greening Program. Christy's training and insight helped see 700 community gardens flourish in New York City. Two decades after her death, her garden (for it is now known as the Liz Christy Garden) has achieved the same official city recognition as Central Park. Along the heaving grey slur that is the Bowery, it offers a tablecloth of serenity; a couple of blue plastic chairs, nestled among bamboo foliage, hint to its trash-heap heritage.

•

Matt would text me while I was sleeping, and I'd wake up to explosions of colour, photographs of the exotic flowers out there, shots of gaudy marigolds and brightly splattered crotons. It was charming to think that while he was on this grand adventure, he was seeing the plants and thinking of me. But I also saw this temporary distance as a bit of a science experiment, a combination of physics and chemistry, what withdrawals would occur without the new familiarity of him.

Since the summer I had come to rely on various crutches to stave off what was arguably my greatest fear: loneliness. Girlfriends; fleeting, fantastical lovers; hasty new friends and old, once-lost ones, leaned on in equal eagerness; family; new, short-term flatmates and, of course, Matt – all of these people had made me feel loved and worthwhile, enabled me to work out how to function again. I'd made myself busy with them to avoid confronting the gulf left behind by Josh's absence, and while I'd become better at being alone over the past few months, there hadn't been many occasions when I had been.

Now, though, I was. Sunday nights and weekday mornings, those drifting, freetime hours when one folds into the habits of another person. Since establishing that I would be leaving the flat at some vague point later in the year, I had begun to seriously contemplate living on my own and in my own space. Even though I knew I'd wind up in somewhere that was far further out of town, much smaller and far less beautiful, I was beginning to see the notion of a place of my own as a pleasing graduation. With Matt away, I'd play out this solitude, remind myself of how lucky I was to have it. It was a more grown-up version of those freewheeling bike rides I'd made years before, conquering this odd new notion of home, attempting to confine and control something wilder and larger than myself; something that had the potential to eat me up if I didn't put up defences, show it what I was made of. I tried not to let the dark, quiet nights sucker me in, but embrace them, even if performatively. Trying to master those depths of aloneness while, at the same time, blindly reaching into them in an attempt to touch their edges.

A week after he left February's weather turned furious. For a full five days, Storm Doris ravaged the country. In the city, we are often too built up, too insulated by our concrete castles, to feel the full pelt of such inclemence. We only witness the metres-high waves and fallen trees in photographs, the trains stop around the commuter belt while we retreat indoors and descend underground, our lives continuing in spite of nature's rages. Not so with Storm Doris. The sky turned a delirious, eerie colour somewhere between pewter and pink. Unfathomably pale and fluffy clouds hovered as the gusts prevailed. I came home and found the balcony door and the bedroom window blown back against their hinges, an aggressive freshness rushing through the central corridor that ran the length of the flat. The wind had blown in, forced doors to bang, pushed aside the barriers between outside and in. I don't know if I'd left them off their latches or if it was the strength of the weather alone, but the air had made itself known: here was something uncontrollable, and we mortals had to sit back and bear witness, relinquish what little command we ever believed we had.

It took me back to the first night I had spent there. That was windy, too. The constant rain had given way to a whistling gale.

I lay on the mattress we'd put on the floor in lieu of any furniture and fretted as it rattled the Crittall-framed windows against their iron casings, letting in a draught that would bother the curtains that wouldn't be hung for several weeks yet. I willed the strange sounds, the soft wailing and metallic banging, to quieten and take their leave. The wind pushed under the door, near where our heads lay, and I felt so swept up in it, this sense of responsibility. Those noises were so daunting. I worried that this new, grown-up thing, this expensive flat we had signed our names to, would be shattered by nature's forceful gusts. I wanted to stop them, take them out of the equation. This had never been part of the plan.

But then none of this had been, either. As I lay there alone, I listened. The rattling windows no longer bothered me. Some months before I'd clasped the noisiest up with a rubber band, but also I knew what would happen: the gusts would blow – they were travelling around forty-five miles per hour out there, less than half of what battered the rest of the country – and the glass would hold; if it didn't, I would fix it. I pushed the balcony door shut against the wind, heard the dry bamboo catch in it, like paper streamers stuck on an electric fan. I was no longer scared of it; the flat was no longer precious in that way; we had broken the future it had once held for us.

But this went beyond infrastructure. Now, I could see that that was all right, that grand plans would fracture, bend and change. That things wouldn't work out. After weeks of feeling so turbulent and crotchety, I felt clear, calm acceptance for the first time as the cold air of the draught brushed against my face. It wasn't that I knew what lay ahead – I had no concept of where I would live nor how I would afford to live there or when that would happen – but that it was daft to be worried about the wind or any other of the millions of other matters I couldn't change or control. I fell asleep listening to it howl and rattle the windows; I had blown myself out – let it do the same.

I knew, of course, that the plants would have taken a beating. Dawn arrived in an unfussy show of a thing, a rich royal blue, pin-pricked with lights of a city that had been wiped clean. The overwintering geraniums, too dry to be weighed down by wet earth, were among the first to shed their pots in the gale. Their rootballs and soil were left in lumps at one end of the balcony, their plastic flower pots crumpled behind planters at the other end. It looked more dramatic than it was; plonk them back in pots with a bit of a drink and a pat and they would see another season yet. The heucheras had been ravaged. The riot of maroon and peach leaves, made for catching alight in winter's slow sunsets, had been shrivelled and scalped by the storm. I silently chastised myself for not bringing the trough down to the balcony floor, saving it such a violent deadheading. But these were small losses, really; I didn't have a greenhouse to turn over or splinter; I hadn't laid out baby broad beans in lovingly prepared beds only to see the whole lot be eviscerated. But those bereaved gardeners found positivity among the devastation – that newly exposed land, they reasoned, saved them a few hours' weeding. They had been meaning to replace the polytunnel anyway. These more seasoned growers listened to nature, realised who was in charge and worked out what to do in its wake. They made good with what had been left behind.

The hyacinths had been bent and snapped on their stems but had managed to stay on the balcony. I severed the remaining, life-holding green sinews and took them inside, popped them in a jam jar and placed it on the bedroom windowsill. Their flowers, dark as oil in a puddle, gained a new gleam in the clear afternoon light that followed. I woke up to the sight and smell of them, cheered as much by the fact they had survived the storm as I was that I had saved them.

MARCH

THE SEEDLINGS HAD BEEN JOSTLING for space on the windowsill. It was a perfect germinating spot: plenty of light from the large, iron-framed windows, warmth from the radiator below them. The windowsill was made of red tile, so there was no need to be precious about damp soil warping wood or damaging paintwork. But most of all, I could watch them grow. This was the window next to where the table lay, adjacent to the balcony door. And so, as I ate breakfast in the newly light mornings or wrote in the evenings, I could monitor these infants' progress: from first curve of green stem breaching the soil, pushed from the seed just below, to first true leaves. With each passing day they would gain vigour and strength, their stems would thicken. To come home after a sunny day would be to find them leaning eagerly towards the window, and so I would turn the trays, pots and party cups (I'd exhausted my supply of leftover plastic flowerpots, which would clutter up the nooks on the balcony at all other times of year) around, so that the next day they would lean in the other direction, in the hope that straightness would emerge.

I'd sown the seeds in the weeks before. There's an inexplicable kind of panicky competition between gardeners about when to sow; the keen will put sweet pea to potting soil in November (after, most likely, a soak in damp paper towel and a Tupperware for a couple of nights), others will plant straight into the earth without much fuss in May. There are those who sow to get ahead while the days are still short and dark, and as many who argue that the little light-deprived weaklings that emerge aren't worth the preparation when you can do it two months later and the resulting plants will catch up just as quickly. Then, of course, there are those who just skip the whole affair and stick nice, healthy plugs in come June. Most people do a combination of all three.

But I had been fidgety in January and, as much as plant vigour, the reason to sow in the depths of winter is because it offers some kind of gardening when one has exhausted trawling through the seed catalogues. Tomatoes and chillies, as largely indoorsy things, are well worth starting at the beginning of the year. And these, along with some sweet peas, now comprised a small jungle of toddler plants that offered noticeable growth to inspect with every new breakfast.

I nabbed a fruit box from the corner shop and carefully lined them up inside it, damp bottoms darkening the cardboard, green tufts swaying above the corrugated edge. The seedlings were going to my downstairs neighbour for two weeks, because I was leaving the flat and didn't trust their care to the lodger.

I wasn't leaving for another home or another's bed. I was leaving for a holiday: two weeks in Japan's wilder reaches. It was an ambitious adventure, and yet not one I'd wanted to take again.

In recent years I'd scrimped and saved to make a couple of trips there with Josh; it was a place that he loved for its quirks and culture, and I had always been a happy travel companion. Over time, I had come to fall for the way nature worked there, how it was folded into the meticulously organised infrastructure that man had made, and that helped me to adopt his fondness for it. But even as an addictive, sprawling place, a third trip felt unnecessary to me. There is an infectious completionism to the way millennials travel; we visit one country for a short while and consider it done, ticked off. People couldn't understand why we kept going back to Japan and, when we'd booked the tickets some months before, neither could I. The stretches of Latin America or Africa were unexplored by my small English feet. Travelling was something I did a lot but squirreled away money to manage: exploring the world was yet another thing millennials were expected to do. Still, we booked the tickets to Tokyo. It was an attempt at making something better, giving us something to look forward to. But it came to be a holdfast that let us down, the staple connected to a fantasy future that ended up coming loose. Now, determined not to let the flights pass by, I was going alone.

I was panicked by the prospect. On paper, solo travelling was an aspiration, presented as the ultimate in millennial freedom: time away from colleagues and loved ones to reconnect with ourselves, meet other people and gain the 'authentic' experiences beyond the strictures of our comfort zone. It came with images of freckly free spirits wielding backpacks and wholesome facial expressions. I wasn't convinced by any of it, just sheer fear at the notion of spending two weeks with only my own thoughts for company. I couldn't remember the last time I had been by myself for an entire day, let alone a fortnight. I went into planning overdrive, reached out to seek advice from friends who went away by themselves a lot (hang out in hostel common rooms and book cookery courses, both of which sounded less appealing than solitude) and attempted convoluted plans with long-lost friends who were in the region at similar times. I made complicated bookings in remote guesthouses and drew up detailed itineraries in an attempt to quell my anxiety with written order. Meanwhile, I pretended to friends and family that I was raring to go, couldn't wait to get a break away from London. The flights approached with an unmistakable dread.

The direct flight from Heathrow to Narita Airport is about thirteen hours, and the seat next to me, which should have been filled with the long expanse of Josh, was empty. I slept for much of it, drifting off in the unanticipated abundance of cabin space. When I woke, I pulled out the itineraries I'd made and printed off, read them like scripture, over and again on the plane, imagining train stations I was yet to encounter.

But as we neared Narita, the nerves lifted. I looked out the window to see mountains rise up from behind the wing of the plane, found myself saying 'wow' to nobody in particular. With the ground came a more domestic landscape, flattening grey-brown squares of fields that grew ever larger; the odd tiled roof and puff of leafless tree. I felt my veins judder with an anticipation that was more good than bad, grabbed the armrests in excitement. I remembered why I was there: not because of convenience, or of not letting things go to waste, but because I find giddiness in new places, in seeing the minutiae of life on the other side of the world. And I had two weeks to myself to find it.

Beyond London, Greenwich Mean Time and all the trappings of commute and office hours and bedtime that come with it, I became answerable only to me. I had long been hungry for more hours – to spend with Josh; then, in those first shallow-breathed weekends after the break-up, out of the flat and into more time, once I got comfortable in it again, to garden and faff about. I had overfilled the hole left by my past relationship with lifestuff: projects and books and friendships and plants. I had become extremely good at being efficient, at first out of survival and then from habit. And when I didn't have reason or need to be so – a spare evening in the diary, a gap in the day – I would fret over it. In endlessly pushing to do more, to have more and be more, I had made myself incapable of relaxing. It felt as if every spare hour was one wasted: there was always another Netflix series to be au fait with, a zeitgeisty article to read or an album to listen to.

I started off in Kanazawa, a small city near the middle of the northern coast of Honshu, Japan. Here there were none of the crutches I had come to rely on to avoid the quiet times. The accidentally alone times. My friends weren't here; I was away from the office, on a different time zone to everyone's social media posts and removed from the plants. Here I was merely to be. I had brought four paperbacks and a vague determination to write. I may have made grand statements about wanting to slow down, but I had little idea as to how. And so I didn't.

Having reached Kanazawa on Shinkansen after landing, I'd

showered only to head back out again into the dark, pacing the streets. I climbed the external stairs of a block of flats to get a view over the skittering lights of the city, remnants of sunset faint lines against the sky, and stumbled hungrily across one of the gardens the city is famous for, Gyokusen-en, lit up and gaudy as a bauble. The Japanese go to Kanazawa for the gardens, and I cycled there but found watching other people enjoy them – with selfie sticks and fancy outfits – more interesting than the horticulture involved. Japanese gardening revolves around control and focus; in a country where nature proves its power with the turbulence of earthquakes and tsunamis, to garden is to seek perfection. I watched small armies of crouched-over women weed the moss lawns by hand, wondered what I was missing. It all felt so static; I struggled to connect. Later that afternoon, I ran the bike along the cobbled pavements by the river (cycling on them in Japan is far less illegal than it is in England) and turned off into the hills beyond the city's wood-panelled old town. Here things were more untamed, more domestic. The heavy bike struggled with the steep narrow paths, so I parked it by a lamppost and continued on foot. Stray bamboo leaves pushed at my sides as I walked through the streets, gradually losing my breath with the incline. It was still cool, the year young, and most of the trees were bare. But there was plenty of life to be seen, including an old man perched on top of his greenhouse, making amends. Beneath him, growth pushed against the panes of glass, a box of mist and green.

I woke the next morning feeling vaguely triumphant. The cross-country trip to Mount Kōya lay ahead of me like a quilt, comfortably conquestable, no longer the preposterous fiction it had seemed a week earlier. The train from Kanazawa to Osaka, in the crook of Honshu's southern coast, was a near-three-hour glide through wide stretches of river and mountain. From the industrial grit of Osaka, the Nankai Express steadily ascended the hills. The houses started to shrink, the roofs gained slope and tile and the light at the end of each cliff-side tunnel only offered more charming scenes. Lone station masters in epaulettes and hats waved white-gloved hands as we passed small wooden stations; elderly gardeners, tending to allotments and greenhouses, looked up at the train as if it didn't pass by several times a day. And then, when the forest took over, it was a cable car that carved through those steep inclines of snow-dusted bamboo and looming cedar. I got out and greeted the chilled air, found a bus and followed a hand-drawn map, sent over email, down a lane to find Kongō Sanmai-in, the temple where I was to spend the next two nights.

The night before, the monks had emailed to tell me there wouldn't be any dinner provided. 'There is no material', read the short message, before a full stop. One café was open, and as I ate there I stared at the abyss of time ahead of me. It was five p.m., the time when, out of season, Mount Kōya closed down. This was less a village than a religious settlement, where around 120 temples have been built over the centuries in the geological bowl that separates eight mountain summits in the Kii mountain range. In the summer, when it is warm, Mount Kōya lures in thousands of tourists. But this was early March and, as the few pairs of shoes at the entrance to my lodgings suggested, I was one of a handful of visitors.

I stocked up on sweet snacks in the mini-market next door out of fear of evening hunger – there wouldn't be food until breakfast after the morning service the next day – and headed west, towards the fading light and past enormous, imposing temples that rose ethereally from the towering cedars. They were large enough to seem almost godlike in themselves, dwarfing the one man at prayer on the steps, his brown robes camouflaging him in the twilight.

In the imminent dusk, the grey and the brown of the woodland and temples collided, becoming a background for the white of the O-mikuji – little strips of paper bearing fortunes – that had been tied onto the trees and boards of metal wires standing in the ground. There was no breeze, and these hundreds of tiny little flags lay still. I have never tied an O-mikuji, but they are a common sight across Japan, most often near Buddhist temples and Shinto shrines. You buy them from the stalls nearby, or sometimes put money in an honesty box. The piece of paper selected at random holds either a blessing or a curse and a fortune, rising from marriage proposal to illness. All life is catered for on these little tokens, from the banality of market speculation and business transactions to the poetry of people being longed or waited for. To take an O-mikuji is to indulge in a lucky dip of life. Tradition dictates that the fortune can be mediated by attaching the O-mikuji to a nearby tree, which will either exemplify or negate the small paper prophecy – good ones are tied to *Matsu*, pine trees, and bad to *Sugi*, cedar trees. Both words have double meanings: *Matsu* 'to wait' and *Sugi* 'to pass'. Good luck, therefore, can wait, while bad luck can pass. Eventually, all of the fortunes are

went down as I wallowed in unwelcome loneliness and shame-riddled boredom. Whatever I had imagined I would find up in this pine-needle-covered temple town of pin-drop quiet, the hoary insight encouraged by *Eat, Pray, Love* narratives or the vagaries of 'spirituality' that people gain from gap years and exotic sabbaticals, it did not lie with the soaked sweet grains in the bottom of my cup. And so I paid up and left, back to the room in the monastery, where the mattress had been unrolled on the tatami mats. The shower swiftly ran cold, and I was led to soggily investigate the maze of dark corridors beyond before I found a private onsen in what I expected to be the toilet. I let myself down into the steep heat of the water, willing sleep to come to stave off the remnants of the waking night.

People stay in Mount Kōya's monasteries because there are considerably more of them than hotels or hostels. But there is also the allure of joining the resident monks in their ceremonies, which start shortly after dawn and continue throughout the day. And so I went, too, waking at six-thirty a.m. after a fitful night of forest-filled dreams and swiftly dressing in the same practical, comfortable layers that I had worn, in different orders, since the start of the trip. The whole of Kongō Sanmai-in was cold; snow had fallen overnight, so to push apart the sliding doors that opened onto the thinly covered porch next to the temple garden was to let in a gulping draught. I began an idle day of ceremony viewing, watching four monks undertake rituals of a significance I couldn't understand, finding less meditation than my mind wandering into riddles I would never solve. Their chanting was at once alien and familiar, the loops and bass notes of their quartet of voices reminding me of dance music. Hours later, I would find myself at another ceremony, in another, grander temple that held air heavy with incense and the leaden weight of somewhere that is very dark and very holy. When the chanting began it sounded somehow purer than before, and I drifted against it, losing grasp of time, knowing only that my socked feet had grown numb against the floor.

This temple had been in Okunoin, the cemetery famed in Japan for its size – more than 200,000 graves – which, with the temples, contributes to the sanctity of Kōya-san. I'd visited cemeteries on previous trips to Japan and always found them intriguing places. Quiet, of course, but filled with a kind of charming familiarity with the dead that we seem incapable of in England. I liked seeing the hobbies of the departed (books, karaoke microphones) etched onto their tombstones, found endearment in the cans of coffee and sweets that were left as offerings. Okunoin had the added benefit of age and nature; the graves smother the floor of a forest. While there was a more open section for the 300 or so corporate 'tombs' built to remember employees of corporations such as Nissan, Panasonic and Kirin, producer of beer and tea, the bulk of the stones were laid beneath the trees: cedar, fir, red pine, southern hemlock, the Japanese umbrella pine and Hinoki cypress, famed for its intoxicating, earthy scent.

This forest became protected almost by default in the early nineteenth century, when a ban was put in place against these six trees being used for anything other than temple-building. It's difficult to imagine a place so sacrosanct bearing witness to anything other than piety, but Kōya-san has seen all manner of devastation, from war to fire. After each deluge, the people who chose to settle there for peace have cleared the debris and begun again. Many of the temples here are not as old as they seem, meaning the surrounding forestry has not only outlived them, but also been planted with intention: a means of creating something anew.

There is a clear path through the cemetery but, tree-like, it has many branches that started off legibly enough before whittling into the forest floor. There is no sense of prohibition here; I was allowed to wander where I wished, it seemed, as long as I was quiet and cautious and respectful. And so I'd find myself up a hillside or some long-worn steps to find smaller, tucked-away graves more deeply surrounded by trees, the softness underfoot of dropped conifer branches, bracken and moss deepening with each step. Other times, these pathways would lead me to the surreal and unexpected: a broad, windswept plain of post-winter yellow scrubland, as if I hadn't, seconds earlier, been in the throes of woodland at all.

Focusing on that unspoken, near-primal mission of discovery took me out of my panicky boredom and loneliness and into a more settled, contemplative state. It was impossible not to acknowledge the passing of time, even while being lost in the present – I could have been there hours or merely on a detour that lasted seconds, it was difficult to tell – because nature made a living history before me. Into the further, more inaccessible reaches of the cemetery the stones blended into the forest under a blanket of moss; I wondered who had laid them and whether their descendents were still alive, who mourned the people whose markers could no longer be read. The Buddhist markers, traditionally composed of five stones piled upon one another, went back hundreds of metres, forming their own architecture beneath the undergrowth.

Frequently, the green and the grey would be interrupted by a dash of red cloth. Some were vibrant, but more often than not they were faded to dusty pink or pale orange: the garments of Jizo statues, the poignancy of which I never got used to. Jizo is the Japanese version of a Buddhist god who protects the vulnerable and the weak, but also women and travellers and, particularly, children who died before their parents, whether in the womb or out of it. According to Japanese Buddhist beliefs, those who die in childhood or even before birth are destined for purgatory, because they didn't have enough time to build up good karma during their short lives. But Jizo, who has long sleeves, smuggles these children's souls to paradise. He's a reassuring benefactor for grieving parents and women who have lost children in miscarriage, infancy or abortion. Jizo statues stand out in a cemetery. There are a lot of them, but they are small and often in the shape of humans – monks or peaceful-looking infants. Most distinctive, though, are the red bibs, hats and knitted cardigans or jackets that the Jizo statues wear, dressed by people who are mourning children not large enough to have their own graves.

The Jizo statues may have offered the clearest clue, translation even, to the lives of those who were remembered at the cemetery, but even though I was conscious of not understanding so much – the Japanese characters engraved on stones, the reverberations of the distant gongs and chants – there was something in this space that offered me a stillness I had struggled to find elsewhere. This was transportative ground; it felt different from the rigidity and order of the temple gardens, which held raked stones and clipped trees. Even the weather seemed ethereal – snow fell from clear skies, drifting in flakes from the gently swaying tree canopies. Then clouds would appear, seemingly only to be burned through by sunlight that caught the cusp of foliage before dancing on the ground beneath my feet. The weight of memory and loss was undeniable, it seemed to distil the air into something that pressed upon my very being, beyond emotion into the realms of the physical. But it offered a kind of relief, too. After days of waiting for an arrival – of gravity, of awakening, of feeling changed, somehow – it had come. My solitary travels, which had left me restless and bored, finally granted me a sense of significance. It rested on my shoulders like a slightly too big coat, wrapped me up in it. I was encased in something that I had only previously been able to witness in drifts and flurries from afar; it felt like resolution.

I got back mid afternoon and found my room warm, my bed made and an even bigger thermos of tea left in the wake of the smaller one I had exhausted the evening before. Snow started to fall outside again, and I sat on one of the wicker stools in the porch bit of the room to watch the flakes settle on the stones of the garden beyond. The night before I had thought mostly of escape, of abandoning the temple and the solitude of Kōya-san and leaping on the first cable car, then first mountain train, back to Osaka, a city teeming with life and batter-based delicacies that I've always been fond of. Kōya-san was just too quiet; I felt claustrophobic surrounded by the nothingness of it all. But here, sitting in the little conservatory I had shut the door against before, I realised what a luxury I had afforded myself, to be a bit bored. To get used to the surroundings, appreciate their extraordinariness, listen clearly to the noises in my head because there was nobody else to speak to, no internet to connect over, only mysterious ritual and great feeling etched into the landscape.

I was drawn back to the cemetery that evening. The snow was persistent – enough to cover the pavements, whisper faint crunches with every step. In the dark it was transformative, more a feat of magic than meteorology, as if it had fallen only once the temples had shut their heavy wooden doors, flakes falling safely in the knowledge that most humans were hidden away. With nightfall, the path had been lit, electric bulbs in lanterns cast the yellow path and my route was more conservative than the scrambling adventure I'd riddled across the place a few hours earlier. But the cemetery had changed nonetheless; I felt as if I were seeing a different place even while retracing those most recent of steps. The statues stood in their own darkness, granted a new mysticism by the snow resting on their heads and noses. Like that landscape I, too, was changed. I was calmer, more willing to be less in control, less determined to plan than to wander. I had given over some of that restlessness, that need to change and challenge. Instead, I was beginning to allow things to be.

I swapped one mountain range for another, Gifu, where I was staying in a whippet-thin string of a village named Magome; the kind of place that clings to a hillside and swiftly runs out of weather-worn houses, leaving only expanses of hard and barren beauty. Heavy, brindled clouds inched over the mountain peaks that crowned the surrounding forests of cedar and bamboo, occasionally letting in glints of the last light of the day, which struck the flattened surface of battered yellow grass, weak from a cold winter and relentless wind. Even in the upper reaches of the village there were neat vegetable patches. Green shoots emerged from neatly laid black liners, brassicas sat proudly, perhaps knowing they would taste better for last night's snowfall. Bamboo canes lined neat furrows of earth and the red capes of scarecrows blew in the wind. There were allotments and green netting over metal frames, hinting at the crops to come later in the year when it is warmer. In the small hilltop cemetery I struggled to identify fat-budded bulbs, about to bloom any day.

Conversations in the dining hall opposite the guesthouse suggested that I was alone in staying here for two nights. Most people were in Magome only for a few hours, leaving early in the morning to walk the Nakasendo Trail – a 400-year-old former feudal highway – and staying in Tsumago, the next, and a far prettier, village along the way. I felt the isolation begin to bite again, conscious that I was wrongly idling with my time. But after dinner I slipped into the routine that I had now become dependent on, of inching myself into the deep, hot waters of the onsen, wrapping myself up in the yukata – a lightweight, ankle-length kimono provided by most guesthouses – and settling down to read on the bed that had been made up on the floor. I'd now spent a week in traditional Japanese houses, and I'd become comfortable with their chill, the gentle shudder of the sliding doors, the way they invited hush and contemplation. I'd started to shrug off the trappings of city life; I'd barely delved deeper than the top two inches of my suitcase, happily throwing on the same clothes day after day. I'd eschewed make-up and tied up my hair; mirrors had stopped being a concern – there were few people here to see me.

It wasn't so much that I was alone with my thoughts as just alone, forced to confront myself without the pleasant distractions of having others around and lean into the luxury of doing exactly as I pleased, rather than what I felt I should be doing. I was pushed into realising that I liked to take my own pace – I was frequently overtaken by better-prepared walkers – and explore the shrines and gardens and neighbourhoods I passed on the walk. I learned that I had brought the wrong shoes – everyone else was in walking boots, I was in Nike Air Max – but that I picked myself up when I slipped in the snow. I began to understand that I'd spent so much time thinking about the intricacies of this trip – the long journeys printed out on paper and the far-flung guesthouses – because I hadn't wanted to face the reality of it. That this was meant to have been something else, something shared, something for two that was now entirely something for one. That, really, it didn't matter where I went, the miles I had racked up or the extent of the isolation I had gained, because I was always going to be here by myself, and that was a greater challenge than whichever mountains I had to navigate around. This solitude wasn't something I could prepare for with reading and research, because I'd never subjected myself to it. After months of wrangling with admin and practicalities, so many of my feelings had been in relation to other people – Josh, or Matt, or my friends. I'd never given myself the space to be submerged into what I felt without them.

It wasn't that there was a grand light-bulb moment for any of this. The Nakasendo Trail was undeniably beautiful – and remote: the only signs of life came from the silent smoke that rose softly from chimneys, or meeting the occasional dog and, in a combination of the two, a welcoming man who lived with his excellent hound in a low, wooden building, and served me green tea from a pan hanging over an open fire. But I took in the snow-covered bamboo forests and constant thrum of the nearby river for what they were, rather than a place of great enlightenment. My realisation had been a slower, quieter thing, more present in its absence than anything else. I arrived into a sun-bathed Tokyo emerging from dreams as shifting and as light as tracing paper, and felt consciousness stir as the city, now familiar, materialised. I grasped for my bags, for my ticket, as I did the space between meaning and reality. My hair still smelled of woodsmoke.

•

We studied Japan in Geography lessons when I was fourteen. I remember its dated depiction in well-thumbed textbooks, how, even then, it looked like the future. We were told about the superfast trains and earthquake-defying architecture, how everyone lived on top of one another in cities, in tiny little flats. It all came out like a cautionary tale: we were taught about the exhausting work culture, the expectation of company loyalty and long hours. We were told that young people were put under so much pressure to achieve things and conform to expectation that they found it too much to bear, that Japan was plagued with a chronically high suicide rate. It wasn't all hyperbole: Japan's suicide rate is higher than that of the UK, albeit considerably lower than that of Russia. High-profile cases of *karoshi*, or death by overwork, have seen renewed pleas made to the Japanese government in recent years to change the lethal long-hours culture; those who have died are invariably in their twenties. Japanese millennials are infamously avoiding sex and physical relationships, due to a smorgasbord of reasons including technology, unemployment and income.

What's resonant about this, though, is how we grew up to be part of a society that doesn't look so different from that painted in near-dystopian hues when we were teenagers. We have increasingly moved to cities – choosing amenity and more unstable housing over the safe sprawl of the suburbs – in our twenties since the 1980s. Overworking has become the millennial norm. In 2019, the World Health Organization defined burnout, induced by 'chronic workplace stress', as a serious health issue. We have grown up against billboards that tell us we're not 'beach body ready', to be left to find connection through dating apps. The headlines say we're sexless, but economists suggest it's more complicated than that – we're so risk-averse, so riddled by anxiety, that it's just more appealing to stay in than go out on a date.

But it is difficult to play compare and contrast – Japan may have freed itself from isolation in the mid-nineteenth century but there is much of it that remains at a cultural remove from a bystander or tourist. Still, I had come to find Tokyo a comforting place over the years. I liked the neat order of it, how people line up to get on the subway, rather than crowd around the doors. I liked the endless rows of pastel tiles that grid the city's buildings and public toilets. It is a place well set up for solitude, there is no taboo about eating alone and, even as a *gaijin* – or foreigner – I felt happily swept up in the well-ordered millions of bodies that gave the city its lifeblood.

But Tokyo was also a place steeped in meaning for Josh and me. We may have fallen in love in London and lived in it together, but London was a city that I had come to know on my own terms. Its cycle routes were etched in my psyche, my geography of it built from job interviews and dinners and drinks and laughter, of lost friends and making new ones and looking longingly at people having fun in pubs when I was passing by alone. London was a city that I had worked hard to love but was firmly entrenched in; a place founded on the feeble-salaried graft of my teenage ambition and early-twenties survival instinct. Of infuriating commuter-hour supermarket queues and blissful early summer riverside Pimm's and hours and hours and hours spent on buses, thinking about how I would get to that next job, that next flat, that next club night, that next catch-up with friends.

Josh and I roamed London, but we also made ourselves a pocket of it that somehow existed on its own orbit. Not in, but around the city that had propped me up and challenged me for so many years. The past nine months had been an exercise in birthing a whole new London, one that housed the nature brave enough to thrive there, with walls that sprouted buddleja and wildflowers that grew from rubble. A London whose soil I found such solace in digging.

But Tokyo had always been a place made of the adventures Josh and I had shared, the language we'd grown and the habits we had developed. For me, kerbside vending machines and talking toilets and enormous, heaving zebra crossings were the makings of inexplicable, undeniable romance, because they were things we had found together. And I indulged those memories. For the first few days I spent in Tokyo Josh and I were communicating like friends for the first time in weeks, months. We chatted amiably over time zones and thousands of miles of Wi-Fi. I purposefully retraced steps we had taken, followed his tips, went to cafés and shops that we had loved, relished reliving what we had once had, except now I was all by myself.

After months of feeling awkwardness or pain over the memories that were conjured several times a day by the most banal things, I allowed myself to have them. It renewed my gratitude for his best attributes: how much he cared about unearthing hidden treasures, how meticulous he was about the process of it, what joy he found in the smallest things, his delight in somebody else's attention to detail. And by doing that, by allowing the good for just that – good, rather than missing or absent or sadness or loss – I was able to appreciate what we had made as both splendid but also gone. And, like sticky chemicals washing over photography paper and waiting to see what shone through, this was a process that exposed how I wanted to live my life differently. How I occupied Tokyo alone could be how I lived my life alone.

Some friends had happened to be in town and they'd wanted a cocktail at the Park Hyatt bar, to take in the *Lost in Translation* tourist spot as the sun went down. There wasn't a showstopping sunset that evening; instead, illumination came from the ground upwards, the red warning lights (for the benefits of helicopters and passing aircraft) atop Tokyo's many high-rise buildings flashing in their own rhythms, cutting through the dusk like breath. It was a sight I had admired in other less fancy hotels, and always in Josh's company. And here, on the fifty-second floor, in a grasped few seconds of isolation while the girls sat on their phones making use of the Wi-Fi, I looked out at those lights and realised that I was finally bidding goodbye to all that. To Josh, and what we had. I was granting myself permission to fully move on, release myself from the holds of it and place the memories and meaning of what we had somewhere else, somewhere firmly in my past.

After that drink I started to claim Tokyo for myself. I inhaled into the isolation I'd been so initially fearful of, found contentment in the hush. I comfortably let one hour slip into another, giving myself the option to wander or nap when I felt like it. The days were made not of frenetic sightseeing or things to tick off a list, but a kind of happy idleness that, a week earlier, had felt impossible. I was motivated not by thoughts of what I should be doing – what was good for Instagram or according to the guidebook – but only by the desire to do exactly as I wished. I left the print-outs and plans, now tatty and soft with folds, in the front pocket of my suitcase. I slept in, allowing the gradually lifting light to wake me up. The skies were rarely clear blue, but rather an endless, cottony daub of grey, which bathed Tokyo's concrete in a light as soft and comforting as a well-rumpled duvet. I developed my own rituals: searching on Google Maps for spaces kind of renewal. I fell for the narrow gaps that run between each box-like, pastel-coloured house (a Japanese quirk that helps prevent earthquake damage) and were, in the best cases, filled with polystyrene boxes or beer crates teeming with daisies or the fluttering, geometric leaves of *Oxalis triangularis*. I stopped to admire whole walls tiled in little grey squares, a utilitarian backdrop to a generous spray of ruffles of pink blossom. In Kōenji, I walked past an immaculate bonsai shop, catching a glimpse of the proprietor – a man in a traditional hakama sitting at the back – before being distracted by the business next door, in which a young woman dressed as a toddler organised clothing in primary colours and racks of vintage toys to a jerking soundtrack of ska.

Tokyo's traditional gardens had always left me slightly cold. I lacked the practice necessary to gain meaning from the arid monotony of the Zen gardens and found the formality of those I'd visited in the historic temples of Kyoto and Kanazawa, even Kōya-san, stifling somehow. Their art lies in their constancy; it is a very different method of gardening from the weather-beaten dance we perform in the English garden, where seasons will shift depending on the sunshine. But my ambles took me to Tokyo's domestic gardens, where people grow things much like they do at home – because they like them, and they know how to look after them, or maybe because they are colourful. Lust for life tops artistry. Palm trees flanked blocks of flats in Bunkyō, where the streets were rich with the smell of verbena; the Barbie-pink shock of quince flowers shot out at an angle from the middle of a roadside hedge of box. The creamy elegance of a magnolia cut through the grey of an empty billboard looming above a car park.

I seized upon perfect pink anemones in window boxes and triumphant maidenhair ferns in pots, their fronds rippling in the breeze like hair under a dryer.

Shibuya is famous for shopping but alongside those storefronts are rows of upside-down plastic bottles shoved in pots containing saplings and shrubs, an unruly explosion of green on a pavement. What determination, what persistence they had to be there. Both the humans who thought to grow trees in pots on a pavement, and the plants that thrived once they had been put there. I thought about the rituals they must invoke, in their watering, in those plastic bottles, and how just by being there, in one of Tokyo's busiest neighbourhoods, this little container garden invited good grace and patience from those who had to step carefully around it. On my last afternoon in Japan, I was a few stops away in the scruffier neighbourhood of Shimokitazawa. Days of everyday wandering had given the once-different things a veneer of normality. I passed by vending machines and under overhead wires without stopping to breathe in this new vista. But a rare bit of graffitti held my gaze. Next to a pink wall surrounded by ivy, on the ridges of a corrugated iron wall, the English words called out to me after a fortnight of being blinded by kanji: TOKYO IS YOURS.

•

Matt's flight landed a few hours after mine, and he headed straight over from the airport. We'd kept intermittent contact while we had been away; I had wanted to grant him his own time and space away from London life while he was in India, and in turn

I had relished the space of my own company. But I opened the front doors of the building to find him bundling in, bending me backwards with a Hollywood kiss while still clutching onto his dusty suitcase. We were all of a scuffle, no room for words in a reunion that was puppyish, rather than passionate, in the sheer joy of being together again. I was taken aback by my delight in it. My readiness for it. I was happy to be here, being made a fuss of under the fluorescent lighting of the hallway. That tension, the feeling of being torn between love for Matt and a sense of betrayal to Josh, of being so steeped in feeling but far too quickly, had deserted me. We were just being, it was as simple as that. We had granted the next day to one other, to unpack and get sorted but also to get over jet lag and snoozily attempt some kind of reintroduction to the city we met in, but we failed to leave the house; it was foul outside, so we just lay under the covers, listened to Paul Simon and traded stories from our adventures.

I settled back into London a few days before the spring equinox. Technically, it was still winter, but winter had chosen to go out fabulously. The previous week, I had gathered, had been warm, and I returned to find the balcony as full of bustle and bluster as a ship in full gale. The white hyacinths, bought from Columbia Road Market six months earlier and fretted about ever since, had erupted in a drunken riot of rich scent and piercing white on the balcony's eastern end, some so heavy their flowers dragged their stems into curves, drooping over the cushion of artemisia beneath. While I'd been away the tight, hard little buds of the camellia had opened into a gasp of layered petals; they greeted me at the balcony door as I walked out, took in a London about to bloom.

And London in bloom is such a noticeably different thing after a trip to Tokyo. We don't have just one type of cherry blossom – we have dozens. This is largely thanks to the British sakuramori, or cherry guardians, whose obsession with the plants saw them become unlikely, far-flung experts in native Japanese trees. Collingwood Ingram, an aristocratic boy-naturalist-turned-cherry-fanatic grew dozens of varieties in his spacious back garden near Tunbridge Wells, from which he would send and receive cuttings of rare Japanese trees back (they survived the forty-day journey by being stuck into a cut potato) to a country whose soil had been so addled by industrialisation and bloodshed that some varieties had become extinct in a matter of decades. It was his collection that led to the diversity of cherries being grown on British streets, in our sleepy suburbs and on the edges of our post-war council estates. We can't have Hanami here, because there are far more than eight days of cherry blossom. Trees will put on their shows with less rigid order: some starting in the cool early winds of March, others lasting until the first barbecues of May. London's blossom is dark cerise and palest white and all the shades in between, colliding with the grandeur of magnolia and spiky brash yellowness of forsythia and joyful pastel frippery of flamingo willow. London's blossom rarely surrounds canals or offers walkways, but brightens up a bus-based commute or pokes through a fence. Here, the blossom is a beautiful chaos, a diverse jumble of many shades, erupting for those who can enjoy it alongside, rather than in spite of, the surrounding concrete.

•

When I wasn't dreaming about plants, I was waking to the reality of severing myself from the home Josh and I had once made. We had begun the unpleasant process of negotiating ourselves out of the flat, a steady flow of seriously written letters, a whole different kind of heartbreak in stark type on a page. The solicitors' letters would arrive on the doorstep and fill me with dread, both with the fearsome, alien instructions they contained and the fact that they even existed at all. That this was what our relationship – the in-jokes and the stories, the adventures and nesting, the hours of simply hanging out and growing up – had been reduced to. It was just so utterly sad, so very at odds with the playfulness that we had shared. I felt battered by it, not least because I had to participate in it too. The letters couldn't go ignored, the confusing machinations of money and decisions made years before had to be picked over like carrion. Even though we both tried to inject as much care into proceedings as possible, we were now disparate parties, both standing up for our own after spending so many years fighting for the best for one another.

I would scour estate agents' listings and fit in reams of dispiriting viewings around my office hours, rushing there on the bike, phone in hand to navigate new, labyrinthine estates I had not previously known existed. I was wildly fortunate to be one of a tiny percentage of my generation that could afford to buy anything in London at all, something I knew was a rare and special opportunity. But my budget was nevertheless small. I refused to leave south-east London, which, anyway, is the cheapest bit of town, and I refused to look at anywhere that didn't have some kind of outdoor space – even if that was a tiny balcony.

Without a pocket in which to grow and breathe, I would feel like I was suffocating.

The results of such specifications were largely similar: ex-local authority flats, usually built in the Sixties, often partially abandoned after years of renting, frequently damp and pokey and, nearly always, in need of major renovation. The agents rarely bothered to try to give me their spiel – there wasn't much to be said. It takes less than ten minutes for one woman to look at a one-bedroom flat, and so they'd stand there a little impatiently, keeping an eye out for parking inspectors while I tried to assess what light came in when, and just how much needed to be done, and what walls could be knocked down. Nearly always, these places were not right: too expensive, too far-flung, too small and sad. I was prepared to be open-minded, ready to move further out of town and live somewhere less salubrious, the simple maths of one salary paying for somewhere rather than two dictated that from the start. But so few of these homes had even a whiff of hope about them.

As the dreams would attest, I've always been an adept castle-builder. I run away with ideas that explode and unwind from the smallest of things. A throwaway statement about how nice it would be to, say, go on holiday will send me into planning mode, looking up flights and hotels and when we could go. I could never go to a job interview without first idly considering what my life would be like if I were to be successful, what clothes I would wear, the commute I'd make. It's the same impulse that allows me to stand on the balcony in midwinter and paint a mental picture of what it will transform into with the spring and then in the coming summer. It's also why I struggle with unplanned changes, with the fissures of life that crack open on a dull Tuesday afternoon or an otherwise humdrum Saturday morning; I will start trying to conjure possibilities for situations I haven't any grasp of and then realise it's impossible. Not everything can be fixed immediately. I was beginning to learn that, although it was still difficult to accept.

Those first few days of April, then, were to exist between both fantasy and frustration. I'd fall for a beaten-up little flat in a high-rise in Camberwell, overlook its position eight floors up and see only the double-aspect views it boasted. I'd ignore the entire lack of kitchen and dream of the one I would build there instead. I would find myself sketching out little planting designs for the netting-covered balcony and think about which buses I would get on. At the same time, I'd be hurled back to earth by the unfolding and drawn-out procedures that would enable me to move out of the flat Josh and I shared. I found my mind drifting to the gardens I could grow in these abstract spaces, focusing on daft minutiae – where would I, for instance, keep my bike? – rather than the fact the estate agent was refusing to get back to me and the place was too unseemly to get a mortgage.

These tumbling conundrums would play out in my brain on a loop. Mum and Hannah were wildly supportive, always keen to see listings and hear about viewings, applying their creative and open minds to whatever I had dredged up. But I tried to keep it all at a distance from my friends; it felt both gauche and boring, I didn't want to trouble them with it. Matt, too, was kept at a remove. We were so happily unspooling into one another's lives. While I had gained the courage to be my whole messy, fractious self with him – rather than the shimmering, crystalline and supposedly better kind

And there was plenty growing. Prepared gardeners, ones keen for movement and action, get ahead with what they can in the last weeks of winter, but by April the gardens have caught up. Buds start to bloom, seeds germinate into lengthening stems that push through the soil that has kept them incubated. After months of steadily gathering energy below the earth the plants subject themselves to the elements beyond it; they feel the pelt of shower and the gaze of the sun, turn towards it in hunger and fascination. In the community gardens, the tulip bulbs I'd thrown in January ground cold like a last-chance saloon were beginning to flourish in a happy hodge-podge of colour. Early April held both winter's remnants and the promises of spring. The violas, their leggy winter growth recently chopped, were starting to bloom again. Those white cyclamen that had survived since October without their fleshy leaves turning to mush in the rain were throwing final salutations in their soil-destined petals. One brave hyacinth still stood where its fellow bulbs had bloomed and, after drunkenly lumbering over the side of the trough, turned small and brown. I left their leaves to shrivel and turn yellow as spring warmed up. They would stay green for weeks yet, still photosynthesising, gathering sunlight and feeding the bulb below to bloom again the next year. Against this swansong, an anacrusis: the tight buds of ranunculus and camellia – both white, both gorgeous – pushed out and into layers of petals. They looked unfathomably glamorous against the grey of the floor.

I had been turning the windowsill seedlings over breakfast since they had first gained height and now they were growing strong and straight. Now, it was warm enough to start hardening them off, bringing the sweet peas out in the morning to toughen up in the cooler air, taking them back in with nightfall. But there was sowing too. Small, satisfying industry. Pea shoots and nasturtiums, which both have big, brown seeds and germinate like a song, offering strong stems and pleasing little leaves before the week is out if it is unseasonably warm, as that April became. Pea shoots are great; a cut-and-come-again crop that tumbles up and out of even the most crammed container and onto the plate in a wildly pleasing tangle of green. I think of it as something effortless and beautiful, a mouthful of early summer on top of nearly anything that emerges from the kitchen. Those seeds would bear delicious, aphid-free munch far beyond the longest day.

Nasturtiums prefer to take their time, putting out a torrent of lilypad leaves, just astoundingly handsome in the boldness of their structure, the presence of vein across circular green. The buds come in the shape of soft arrowheads, strung to the vines of stems with a tail like a tadpole's, and will, after many seemingly endless, fruitless days, erupt all at once into a handful of orange petals. You can eat the lot: leaf, stem and flower, and they will reward you if the sun is bright and the soil poor enough – for they thrive where there is rubble, where man has paved abandonment – with more fruitful offerings. A nod to Elisabeth Linnaeus (daughter of Carl, the botanist who shut women out of the academic sphere in the eighteenth century) who, at nineteen, established that certain flowers, nasturtiums included, 'flash' as the surrounding light fades. It's to do with how the human eye interprets the contrast between orange and green, but the end result is that a field of nasturtiums can look like a forest fire. Later in the season, when they had been deadheaded as much as I could keep up with, I let these flames grandchildren, died when I was four after three increasingly painful years induced by a string of strokes. My memories of her are few and largely informed by home video and hand-me-down tales. I sort of remember her lying in bed in her nursing home and there is a clearer, earlier memory, too, of my grandfather – who nursed her until he couldn't – instructing me to give her a Cadbury's chocolate button. I offered it by putting it on the plastic tray that sat in front of her chair, sticky from my grubby little hand. It's a testament to both her and the love she instilled in my siblings and parents that she exists in my mind as a far stronger woman. As a grin into the sunlight on a blustery beach deckchair, fixed curls buffered by the wind but frozen in a photograph. Bits of her wardrobe have wound up in mine: the silk skirt that she wore, with unshowy Yorkshire pride, on my parents' wedding day – it is black, with a smattering of small orange flowers, and seemed such a chic risk for a mother-of-the-bride in the late Seventies. I graduated wearing her black wiggle midi-dress from the Fifties, and still put on her taffeta skirt from the same era to go to work. On those days, I wonder what she would have made of it: the granddaughter she never got to know wearing her old, light-faded clothes in a newspaper office – not as a secretary, but as a journalist.

Grandma lives in stories. A loving caricature of her has been woven over the years, from a warp of humour and a weft of memory. I know that she was a stickler for things being clean, tidy and proper; that she kept good company and her mind and body active with games of bridge and golf. I know that she drove ambulances in the war and became a nanny because she loved children, and I know of the softly spoken tragedy of the fact that she only because they are one of our most tamed wildflowers. *Anemone coronaria* are native to the Mediterranean, where they turn hillsides mauve every spring. As with auriculas, they were among the handful of species to become bred as florists' flowers in the late sixteenth century – as old as tulips, although with less of the fever. Other varieties, like the *nemorosa*, are a modest if spreading woodland plant that will pop up in early spring and push happy little pale blue and white flowers through their wide leaves. They open up their petals with the sun, shut them at the prospect of rain or darkness, and resiliently withstand April's gusts to the extent that they've also become known as windflowers.

Because Grandma's first floral love was wildflowers. The daughter of a north Yorkshire station master – his father a station master before him – she, along with Joan, would hunt for wildflowers on the banks of the railways before pressing them and sticking them to paper. They would fashion whole scrapbooks filled with wildflowers, the colours gently fading with time. It was a hobby they'd been raised into: there's a photograph of my great-grandmother, Emily (from whom I gleaned my middle name), sitting with my great-grandfather, Stan, and friends on a woodland bank covered in winter's leftovers, a clutch of modest blooms in her fist and a happy smile on a face that reminds me of my mum. That was in 1914, in the few months between their wedding and the war. Decades later their daughter, my grandmother, was a grown woman. But she could still recall the wildflowers' names, the ragged-robins and the worts. And she would still pick primroses and pack them up in little boxes, resting on damp cotton wool, to send up to my great-grandfather; a package of petals to cheer him up, delicate blooms ripe with nostalgia. The generational habit didn't work the other way quite so much. Grandma encouraged my mother to press flowers, showed her the scrapbooks she had made in girlhood, but Mum admits to not being interested – she thought it was boring, to be presented with those dead things in a book. And she still celebrates things in life more than death, now; it is the birthdays of my late grandparents that she marks, rather than those on which they died.

The wildflowers stayed wild with the arrival of my grandmother's womanhood; a girlish hobby that remained fixed in time, like those stems flattened between paper. No wildflowers were grown in the gardens in which my mother grew up. In the Sixties they were considered weeds and suburban tastes preferred the showy. The frill of a tea rose and the opulent reach of a dahlia were what was expected from the narrow beds that framed immaculate lawns, and so these were what were ushered from the ground. The elegant gangle of angelica and the golden glint of gorse were left to the railway banks of my grandmother's youth.

Certain things get left behind with the graduation of girlhood. Properness descends as expectations mount. It is easier now, of course; time offers girls and women greater opportunities than they have known before. My grandmother was put off by the notion of earning her keep outside of the immaculate home my grandfather's salary paid for; she would cook and clean and shop and keep house. My mother got a job, of course, but has always said she was only given a handful of options – nurse, secretary, teacher. She chose the latter, still rightly takes pride in it. My generation, though, were told we could have it all if we worked hard enough, any career we imagined and, for women, the ability to maintain it regardless of a future motherhood – which is still, somehow, expected of us as the default.

In childhood, I ignored the pink-and-sweet concept of what a girl should be. An unfussy, practical mother and a rough-and-tumble village upbringing allowed little room for a girlishness I wasn't interested in entertaining. The boundaries and divisions of gender expectation remained invisible to me until my early twenties. Girls were no longer outwardly told what was appropriate and what was not; we had seen women sail around the world and win marathons, had learned about Boudicca and Joan of Arc. I knew what ladylike was but never felt too much of a failure for never quite being it. But I was also protected from the barriers I would encounter once I had graduated and, like my grandmother before me, moved to London. Gradually, I bumped up against the injustice of our lot. Of how much we still carried, only silently and invisibly. How womanhood was something we could still shape but only through a million tiny little battles: dozens of expectations challenged; remits pushed against; glass ceilings tapped at; reflections in the mirror to find fault in; wandering hands put up with; street catcalls bustled by; smaller pay cheques being gratefully accepted with apologies. In work, at home, in all of life, I was waking up to womanhood and realising that here there was work to be done that men didn't even think of.

One of the first things I noticed in the immediate absence of Josh was how much I had cleared up after him. The cereal bowls left on the table and the shoes in the hall, the clothes left on the floor and the laundry hanging on the airer. Without discussion, we had slipped into this dance, not just with the flat but years before, when we shared different rooms in one house and mine would be tidy, and his would not. I am not an easy person to live with. Like my grandmother, perhaps, I like things in their places and at home I am restless around mess. But as I settled into living in the flat by myself, without Josh, I found myself waking up to the performance of womanhood I had been living. We had never discussed who would take on what roles, and yet, like my mother before me and hers before her, I had slipped into that of deciding what we ate and when, of making the house look nice before guests turned up and putting things away before bed. Josh picked up other, arguably more essential tasks – he always made sure there was milk and cereal in, he usually put a wash on. But I looked back at the young woman I had been slightly aghast at how I had folded into domesticity as if were my destiny. That this was a role I had donned as if it were an apron and tied it tightly around my waist, without ever really thinking about it.

That had all faded in the wake of the break-up. I hadn't become a slob, but so much of what our relationship had become was tied up with the notions of home and stability, the idea of a future that would tread those paths taken by my mother and grandmother, that would follow what my sister had – of a sensible long-term love, a beautiful home, eventually a ring on my finger and then the rest of it. And now that wasn't on the horizon, not yet, maybe never and certainly not with Josh. I had been released from it brutally and without warning. In the raw space that was left, I had been confronted with the opportunity to re-shape my womanhood, to look past the expectations society had taught me to heap upon myself and make something else instead.

As previous generations of women had before us, we continued to push at the boundaries that had been put on us. Questions of 'having it all' dominated dinner parties and book groups. Fourth wave feminism had bubbled up through a generation of women who had, as girls, idolised the Spice Girls and been given the internet to share their fury, their sense of injustice. That, in turn, sparked millions of debates about what a young woman should be. To settle as a millennial woman – in a nice flat, in a steady relationship – was both seen as a failure and a radical act. We ditched *Good Housekeeping* mags for Instagram and Pinterest, but the focus on cooking and decorating and making lovely our lives remained the same. Such domesticities were still at the heart of the people we thought we should be.

I had been playing with that sense of what kind of woman I was since the first days after the split, alongside the notion of solitude and my lack of familiarity with it, and friends and love and what it was that made a life. But with April came two clearer graduations that caused me to reflect on where I was, where I had come from and where I might go to. Two of my oldest, best friends from different parts of my girlhood were marrying within a week of one another. The first, Anna, had grown up in the same sprawl of dull, meandering villages that I had. Underneath an affinity for the silly and sugary, Anna exhibited a steely, wry determination that I admired and identified with. We found common ground in a shared sense of humour, of being two girls who had escaped our country confines and were bound for more glittering adventures wearing tatty vintage dresses. She and I embarked upon a slightly wonky, no-nonsense version of what a bride and bridesmaid should do. She tried on her wedding dress in my bathroom and I hashed together a hen night that ended up in a club in Peckham, a gaggle of us chatting over pounding beats as the heart-shaped balloon tied to her wrist bobbed above our heads, disco lights bouncing off its pink foil.

Days earlier Matt and I had stood on the side of a hill in Perthshire singing 'Octopus's Garden' as Martin and Emily became man and wife. I still remembered the night they had met, a few weeks into our first year of university, also a decade earlier. Theirs was a relationship of seeming calm and understanding in a time when little played out that way. And even though we – our flat-mates, and probably Emily too – had always known that they would marry, I doubt we could have imagined a day quite so fitting, from the blustery wind that bothered the congregation's well-coiffed hair to the way the hours of freewheeling cèilidh and drumming, big noises in the wilderness, generated too much joyful heat for even the cavernous barn's big stones to contain. And so Matt and I took ourselves outside, gasping for cold air, and found bonfire smoke floating into clear skies and an unambiguous happiness in holding each other's hands.

I watched both these beloved women with whom I had become nearly grown graduate from girlhood with big promises and white dresses and was purely giddy for them, gleeful to be there. But I got caught between memory and reality, too, as if the events unfolding around the confetti and lace were on ciné film, stop-start and charming and somehow removed from the life that I was now living and trying to make sense of. That the women walking down the aisle were both the ones I knew, who shared the kind of vital, impolite intimacies that girls' friendships are built on and fulfilled by, who could be ludicrous and cunning and above all excellent, but also something other, something polished and pristine and near-perfect, just for a day, as they became wives.

It was a gulf I would gravitate into, perhaps because I was pulling away from my girlhood and into something else. That first spring I spent in Newcastle, six months after the unbridled, wide-eyed revelry of Freshers' Week, was a transformative thing. We left it cold in March and came back weeks later in April, the cherry tree outside my window heavy with pink petals, the promise of warmth and the lingering, cerulean skies of summer term. I spent the in-between time at home in the south, somewhat torn between my adolescent boyfriend and the sort-of one I'd taken up north but mostly feeling in flux, knowing that there was change and life in the air and inside of me but with little understanding of how to harness it. I felt it wake up as the ground softened and offered up daffodils, as the trees became smothered in blossom only to push out tender new leaves once it had dropped. Against the cold gusts of spring and in its flashes of sunshine there was something stirring. A decade later, and I was feeling it again, only I was more confident this time.

I was beginning to realise that I had taken the falling-apart of the life I had never envisioned – nor expected – for myself and, after months of grief, of ignorance, of distraction and self-pity, had managed to turn it into something I had defined. I had learned to let some things go – the comforts of certainty, in many aspects of life – and tentatively embraced others, making the most of things that were offered me, valuing the now, rather than getting caught up in the future. It wasn't that I was a better person than when I had been with Josh, nor a wholly different one, merely someone who had learned to grow beyond what was expected of her, not least by herself.

As much as the weddings were gorgeous ceremonies of new stages in Anna and Emily's lives, they also offered space and time to reflect and, in some ways, bid adieu to our girlhoods. I celebrated Emily's in the company of our other old flatmates. We shared a room stuffed with three bunk-beds, as if we were girl guides for the weekend, and funnelled the excitable nervous energy that arrives before a wedding into a brisk walk around the hills that surrounded the venue. It was unseasonably warm, and bright, and we shed jumpers and coats as we roamed through bright gorse and light-trapping woodland, finding tumbledown barns and elaborate fungus under fallen trees, filling the clean air between us with where the years we had been apart had gone. Good, adventurous days out had always brought us all together. Even from the beginning we'd brave the cold of fish and chips on Tynemouth Beach in bright winter, or venture to the banks of the river to help with some marine biology experiment. We didn't bond over sharing shoes or trading make-up tips, but in roly-polys and human pyramids, young women who knew how to puncture pomposity with a ludicrous pun and a *Gotcha!* grin.

I kept feeling it, chasing it, this feeling to embrace nature's raw edges. When the celebrations were over and we bid our goodbyes,

Matt and I flew up to Scotland's northern reaches, to Orkney. And for four hazy, light-long days which would get caught in memory as far longer, we felt the raw and ancient wind on our faces, in our bodies. Rushed into the gusts of it on these rolling and tree-less lands. When night eventually fell, bringing with it cloud and bitter cold, we bundled up against the dark. At times it felt like we were the only people on these tiny, sea mist-laden islands. I heard the wind outside howl, rattle through the chimney of the low stone house we were staying in, but didn't fear it. We allowed our bodies to let new, raging air batter us just as much as we stood against it, tendrils of my hair whipping my flushed cheeks even as I filled my lungs with the stuff.

London, though, had grown hot. Sunshine seared into the city, touching its glass and its brick. Commuters carried their winter coats under their arms and over their shoulders, tourists from warmer climes looked bemused in anoraks. The year's first real blast of heat burned through the light cloud left behind by the dawn, caught us unawares. Late-risers emerged from their houses in optimistic T-shirts, long-covered limbs blinking in the light. That new sunshine landed on the early leaves, the ones still wearing their concertina creases from the nubs they had emerged from. People sought the outside, found a bench or a park, rolled up trouser legs, lay back tentatively on the grass.

Anna and James were marrying in an eighteenth-century mansion that sat, surrounded by green, away from Stoke Newington Church Street. Love and female rebellion is laced into the man-made river that runs past the house: Eliza, the daughter of its second owner, an industrialist named William Crayshaw, wanted to marry the local curate, but her father forbid it. So they waited. Upon Crayshaw's death, in 1834, Eliza – now in her forties – inherited the place and set about marrying her beloved, patient curate, Augustus Clissold. The park and house still bear his name. In 1889, a few years before Brockwell Park opened its gates, Clissold Park welcomed in the masses, having been bought for the public by the Metropolitan Board of Works a few years earlier.

And so, in our frocks and flowers, we followed the trail of Anna's skirt as she walked along the park path to her wedding venue, turning heads of the Londoners who were delighting in the first arrival of warmth after a trudging winter. London transforms under the application of heat. People become lighthearted and determined in their desire to soak it up. Joggers have new, idling pedestrians to contend with, plans are shifted to pubs with gardens and the opportunistic tinkle of the ice cream van's siren floats in the skies.

I found myself bathing in what we had grown up through. Anna and I had galvanised together. We had grown furious at our socially docile adolescences, at society's poor allowance of fortunes, in our early twenties. We had lost our naivety together and honed our abilities to set the world to rights as we shifted into these womanhoods. And that awareness had dragged us out of girlhood, too. As that sunny day diminished into indigo, we sat on the steps of the grand Georgian venue, Heather, Anna, Jamie and I. Shared a cigarette, watched the smoke curl through the air, a happy tangle of head on shoulder and arm around waist, the skeins of the disco gently nudging at our backs. We looked ahead to the summer and I was filled with a sense of its potential finality, of how our lives were all on the cusp of tipping into proper adulthood, the kind where one spends a Saturday being productive in Ikea rather than sleeping it all away.

The summer ahead felt like it might be potentially the last freedom-filled one we had. I wanted to fill myself up with it, with them. I wanted us to catch deep, misaligned tanmarks and find dirt under our fingernails. I wanted our hair to grow long and fair, grace our shoulders. I wanted sweat on our brows, to gulp down water, to be filled with mild confusion and too many options. I wanted one last summer of choking, tear-jerking laughter and arm-spinning dances as the sky turned dark. Grant me this, I begged silently, watching our lives unfold: a last summer of joyful abandon. Free of objects, free of care, full of longing and love.

Our girlhoods were fading, falling into history as we grew inexorably older, took on the mantles of the womanhoods we had witnessed and forming others anew. And while it could have been something to grieve over I chose not to. I chose to remember what they were, cherish what remained and keep a keen eye open for those glimmers of them that would return as we became women, both together and apart. Our lives weren't losing anything as such but changing into something else, something different from – and yet, somehow still informed by – those our mothers and grandmothers had undertaken.

•

As the venue curfew arrived at Anna's wedding, and the guests slowly filtered out through the park and into the streets beyond, pub-hungry, the lights switched on and the bride instructed the remaining few of us to take flowers, take the flowers lest they go to waste. My bouquet survived the hours of dancing, a riotous and sweaty affair that culminated in a schmozzle of day-glo paint, blackened soles and heaving chests, and made it into a pint glass on the windowsill of Matt's bedroom. I woke, eyes blackened by kohl, to delight upon seeing them there, the curl of clematis and lip of sweet pea, bedraggled but beautiful, still. These British-grown, seasonal flowers, they do not last so long. Even when prepared well, stems cut at an angle and conditioned in warm water, their allure lies in their ephemerality: sweet in the moment but not meant to last, and all the more precious for it. It was a Saturday morning, another day even sunnier than the one before, and Heather, Jamie and I decided to go to the park, spring a picnic, while away our hangovers in the unseasonal heat and swap stories of the magic of the day and night before. To seal in the memories before they drifted.

And it seemed to me that this was what we had gained and that my grandmother's generation never had. That I would never have the surety of a husband for life who would provide as long as I kept house. That the future of my job was uncertain, and my home more so. That women of my era had been told we could have it all, only to learn that all was impossible and exhausting and we would need to redefine what we wanted in life, instead. And although I had been pawing for certainty and uprooted by the loss of it, the freedom I had gained in its wake was electrifying now that I had woken up to it. That I would never be bound by the strictures of growing tea roses in my garden but could grow anemones in a bucket on a balcony of a home that would shortly no longer be mine, and cut them to put in a glass – or not. That I would never be told to choose between just three different avenues of work, nor expected to wholly abandon it to bear children – although that would become its own, other battle to fight. But that, if I could weather the storm of uncertainty then its rewards would lie in everything from the joys of a spontaneous picnic on a sunny day or a late-night stroll around Soho to the greater, more resounding ability to shun those things expected from me because I could figure it out by myself.

And none of it might last long. The beauty of an anemone, like the blossom that was drifting off the trees and dressing the pavements pink by the end of the month, was that its petals would only stand up to the wind for so long. After a few days, perhaps a couple of weeks, these would spread and fall, leaving seeds behind. Those, in turn, would be carried by the gusts, land elsewhere and maybe settle and germinate or maybe not, maybe achieve nothing of the sort. But when they did, they would thrive and bloom themselves. And for the chance of that, perhaps all of it was worth it.

•

We were long into the month and it was a different creation from that which had blown in weeks before. The cow parsley had sprigged up alongside kerbs and railway stations in the city's south-east suburbs, dashes of white on swaying stems. It didn't start to get dark until eight o'clock, now. Windows were pushed open on buses. The last of the hellebores, papery and pale, curled over beneath boisterous new foliage, which had filled the trees and the hedgerows as if by green surprise. The city was gradually growing wild, hybrid bluebells, the progeny of garden escapees that had mingled with the English variety native to woodland, cropped up in traffic islands and municipal flowerbeds. Forget-me-nots crept up bits of forgotten wall. Spring was concocting things; pollen had begun to drift from tree to pavement like icing sugar that missed the bowl.

I had worked up the kind of sweat that comes with rushing around in a warmer-than-necessary jacket in another new part of south London. It was late afternoon and Hannah, with my plump nephew – now a sturdy six months old – in the buggy, had accompanied me to view a minuscule garden flat on the outskirts of Brockley. The estate agent was late and had an air of shiftiness about him; there was some shaggy-dog story about knowing the owner and helping him out, perhaps to disguise the fact the place was a studio flat that had been granted an extra wall to carve out a tiny bedroom. There was a garden, it was true, but we chalked it up as a loss and I thanked and left her to get the baby home for a sleep while I rushed to catch one of those small, always slow buses that tend to bumble around the outskirts of town.

The place I was going to had been on the market for a while; in fact I'd seen it listed half a year before, when I'd taken an idle look to see what, if anything, I could afford. It seemed like a beacon of hope then – a generous, if shabby, one-bed with a leafy balcony perched on the edge of Dulwich Golf Course. On the map it was surrounded by green. But the owner had wanted too much for it, only to reduce the price (still steep, still more than I could afford) about a week earlier. It wasn't easy to find, tucked into the corner of an illogically laid-out housing estate, and the agent was waiting by the time I had paced up the hill, smiling lightly at the children playing on the grassy banks that the blocks of flats surrounded. We made the kind of awkward small talk I had become accustomed to, the daft game of cat-and-mouse where the prospective customer must adopt an air of practised apathy about the proceedings. Red door, opening to tatty poster and then, a turn to the left, a wall of new green leaves. The whole flat looked out onto the woods; established oak trees were just coming into leaf, dangling flower flotsam still attached and catching the sun that would be setting a few hours later. The place was tired; it smelled musty and was filled with sad furniture and heavy curtains. Chipped polystyrene ceiling tiles loomed over the bathroom and bedroom. The kitchen was panelled in orange pine and there was a bizarre alcove in the middle of the place that ate up most of it. It was a trek from the bus stop; twenty minutes at best from the nearest Tube. And yet, it felt like potential. It felt like it could be home, that I could be happy here. That to be among the trees was what I needed, to walk in the door and be greeted by the outdoors. That the steadiness I had been seeking – in the plants, their processes, their possibility and willingness to adapt – lay here, in the great oaks that wrapped the place in a woodland cocoon. I acted unfussed to the estate agent, then called my mum and sister, told them they had to come and see it with me again next week.

MAY

If it had not been for those small hours I'm not sure when it would have started. Either my friendship with Heather, or my understanding of the sensory pleasure that having plants around could bring. Of course, the pursuit of them, the active growing of them, that wouldn't begin for years yet. But that first warm night in 2008 was when I felt the full-bodied joy of it, the reason why I'd come back to the plants, find fascination and solace in them, long for them, seek them out, write and dream about them.

We had been thrown together by our mutual friend Katie. Heather and I had seen each other around, oscillated a bit the way girls with similar interests in small places did. We had graced the same dancefloors, our bylines had shared the same inky pages of the student newspaper, but neither of us had the courage to actually forge a friendship. And so Katie set us up. Out we went, a little awkwardly, on a date of sorts. In hindsight, it was fitting that it had been on the dancefloor of one of the city's most cutting-edge clubs, at a party thrown by a fashionable London-based magazine; for the next decade, Heather and I would push ourselves into those sorts of spaces over and over again, our friendship shaped by the sweat and the thrill of them, the unifying shudders of a punishing drop in a dance track and the nonsensical, meandering glee that arrived on the pavement when we had decided it was time to go home. We knew that here was where we both found freedom. We would take each other to parties knowing that, like satellites, we would always keep an eye on one another, protect one another from unwanted attention but let the other fall out of orbit should the right person come along. Our bond was neither clingy nor cool, it rarely entertained jealousy, instead preferring to motor along on a kind of easy, understanding loyalty.

That first night struck early in May, a few weeks before the exams would kick in. Still, it was late enough into spring for Newcastle's skies – always far-reaching, not least in the biting depths of winter – to grasp for darkness with difficulty. Between the spring equinox and the summer solstice, the nights up there never seemed to get truly dark. Rather, they would descend into ever-deeper shades of blue and then back again into chalk as the dawn approached. It would grow cold, but the birds would sing through the night. For the weight of exam anticipation that was meant to becalm the city's student population, the streets were turning celebratory in the warmer days and longer evenings. We abandoned our tights and coats (although some of the hardier girls never wore those anyway); bodies would stretch out on the lawns that rolled away from the library, convincing themselves they were working as hard as they would inside; disposable barbecues would crop up on the grounds around student halls, the chemical-laden smoke filtering through campus.

The cherry blossom on the tree outside my bedroom window now mostly carpeted the lawn beneath its boughs, making way for proud new leaves. I'd started to keep potted herbs on the windowsill, mostly for improving the simple pasta creations I relied on. I had been keeping the window open for a while now; the man I sometimes shared my single bed with preferred the touch of night air on what little of his skin emerged from the cover and I'd become accustomed to the habit, liking the sense of letting in the outdoors, the mechanical sound a light breeze would make against the metal blind.

Heather and I parted ways outside the lawns that surrounded the Civic Centre, tender on pints of cheap beer and the glow that comes with getting to know someone new. It must have been around one a.m., relatively early really but still late enough for the city to seem stilled, somehow. Not many students took to partying once exams had started. My journey home was short and safe – just up through the university's quad and along the well-trodden path to our halls. And I imagine hers was of similar distance to Heaton, to the house she shared with two other girls. But to me the adventure before her seemed perilous (Heather was a couple of years older than me, and she studied in the year above. At the time it felt like she had lived double the life I had) and I was wildly impressed with her nonchalance.

The flat was quiet when I got in, dark and stuffy with it. But I opened my bedroom door to a whole other space: the warm air coming through the window had done so via the basil plant sitting on the sill, and over the evening's hours had filled that small, insipid room with a sweet and botanical headiness that I had never known before. Fresh and beguiling, the air hit my nose, the back of my throat, as if it were a wave that had knocked out my legs. I didn't need to turn the light on; the moon was bright and it almost felt that to alter this delicate balance – of twilight and oxygen and scent – would be to destroy it. So I got into bed, allowed myself to drift in it, let it play with my dreams.

It was not a glamorous variety with which to realise that plants could be like music or taste, an almost sixth sense that could cling onto memory and meaning, but that basil worked its way into my brain. Without it, that first night out with Heather would have just been another night out, one of the hundreds we have embarked upon since. But the impact of that sweet concoction branded the event into my memory.

Were I more poetic, I might have changed it to rosemary, which has been associated with remembrance for centuries, or the gratitude-laden parsley. But rather, it was basil, the one variously associated with extremes of love and hatred, overgrown for impact in a supermarket pot. I've always considered it the plant that made me realise I needed plants around, even if it would take me a while to fall into cultivating them myself.

After that night, I started to sense the other things growing around me. Despite a childhood in the countryside I was fairly plant-blind. I'd picked up a handful that were impossible to stay ignorant of – stinging nettles; dandelions; may trees; cow parsley; stickyweed; daffodils; bluebells; the more easily identifiable trees – but most of the greenery I grew up against became furniture, the backdrop to an adolescence that could be claustrophobic in its rurality. Much as I found the elegant confines of Leazes Park more quickening than the expanse of field-based public footpaths I had come from, so it was the city's nature that unblinkered me. There was a whole new education here, to sit at the top of Heaton Park in late spring and watch the swifts duck and dive in the fading light. I fell for the way the trees would erupt in white blossom, punctuating Jesmond's pavements with prettiness, how the petals would catch the northeasterly winds and I'd find them on my jacket after cycling across town. I'd clock the mounds of rosemary growing in the quadrangle, run my hands over them to release the scent as I passed and think about harvesting a couple of cuttings, as my friend often did, to put on roasted potatoes. To walk in the Dene in May was to be immersed in the smell of wild garlic, which covered those shady banks in tides of broad, gently nodding leaves, bursting into white, star-shaped flowers (wild garlic, or *Allium ursinum*, is often confused with three-cornered leek, or *Allium triquetrum*. Both have the pungent, tell-tale scent and taste of garlic and make equally delicious pesto, but the latter blooms earlier and has triangular flower stalks, like a Toblerone) that would bob over the stream looking back at their own reflection.

I had gravitated towards herbs – even if in plastic pots from the supermarket, doomed to die because this, cramming far too many seedlings in one pot for any of them to thrive without cannibalising each other's light, space and water, is the veritable slumlordery of gardening – because they were the cultivated plants I had grown up knowing, if not how to raise then how to use. My mother had always kept a herb patch into which I had been sent to gather from an early age, often bringing back something that wasn't what

desolate, concreted-over and bin-hoarding things that little was done in them but smoke and, when it was warm enough and we too lazy to find more salubrious land, sit. I wouldn't get access to a garden of my own, that wasn't my parents', for six years. When it arrived, it was an overgrown plot behind one of those rare tall London terraced houses that hadn't yet been turned into flats. We had the run of the place, my four (sometimes five) flatmates and I. Bearing with the house as we did one another, and it with us. Slugs would emerge from the kitchen sink and I fell asleep listening to something scrabbling above my head in the attic for years. We never really minded the damp towels and mildew-covered bathroom, which looked grand because it was large enough for a claw-footed bath to stand proudly in the middle of the room – even if it was always too cold to ever sit in it. We didn't see the shabbiness of it. It was within walking distance of the pub, the clubs and our friends' houses. There was room for all of our bikes to stand in a rack in the corner of the dining room and the rent was less than £400 a month.

We painted the lounge and dining room, made of it what we could on small pay packets. While it was scruffy, it retained its Victorian grandeur – the cornicing on the ceiling carried dust but was undamaged by age, the original shutters framed the tall, single-glazed windows. We wore a lot of jumpers, huddled in the grim, low-ceilinged kitchen and threw the kind of parties where you emerge from one room to find dozens of strangers in another. I still don't know how, when the cellar – rigged up with speakers and fairy lights – somehow flooded with half a foot of water, we weren't all electrocuted. The parties punctuated the propulsion we all lived with; that need to earn and survive, to make our way in a city that didn't seem to want or need us. Emotions would veer from euphoria to despair. Among those dancing on the tables, there would be the couple arguing or someone who got too drunk. More invisibly, the ones going through the motions, trying to douse those persistent anxieties with noise and substance in the early hours of the day.

The garden was less something to tend to than an extension of the dilapidated space we were grateful to call home. Come May, the parties would stretch out into the lawn beyond the cracked concrete that covered the side-return, where beer would soak happily into the lawn and we'd string jars up into the trees, fill them with candles. Josh and I would work to clean the garden up a bit (when one of our flatmates left London on a six-month adventure that turned into a three-year escape, he moved in. It was a sweet, love-filled time), losing hours of late-spring evenings to pulling rampant weeds from once-loved raised beds and clawing junk out of the overgrowth at the back. It was satisfying work, a more muscular relative of the outdoors tidying up that would later become my regular meditation of deadheading on the balcony. Our flatmate picked up a lawnmower on the cheap. It was something we got excited about; there was talk of growing vegetables.

But life got in the way. Once the weather faded, we were drawn back indoors – to screens and pubs and dancefloors. Once those beds were left bare, I didn't think to harness them for anything else. It seemed too ambitious somehow, the expanse of it so large it became intimidating. I didn't know what plants to put in there. Hadn't even yet learned to nurture my own desire to do so, in spite of feeling a sting of joy every time I walked past the spires of hollyhock and foxglove that our neighbour coaxed from plastic tubs in her front yard. Our growing spaces, blessed with good sun and regular rainwater, stayed unbothered by human touch. And so the plants that we had tamed grew back; I didn't even notice them enough to find out what they were.

I did, though, grow herbs – or try to. While the ground offered by the actual garden felt untouchable and wild, I tried to colonise the side-return, which was easily accessible from the kitchen (I don't really remember us much locking the back door, which was madness considering where we lived) and thus felt like the kind of space in which my mother had always grown herbs – a near-grabbable distance from the various worktops of my childhood homes. The difference, though, was that Mum had always the knowledge and good sense to grow her herbs in patches where they would get the requisite shelter and sunshine. And the side-return was shaded, ferociously damp but still in rain shadow and partially enclosed by a broken fence. I put out basil, rosemary and lavender – sun-lovers all – and figured they would thrive, but they were soon stringy and sad, stretching for the light or devoured by slugs. Had I known to plant them in the garden, in those troughs of bare poor soil, who knows what those modest successes may have led to?

Those fated plants had been cycled up Kingsland Road from Columbia Road Flower Market. The market is one of those things well-known to Londoners that has become increasingly popular with tourists; to go any later than eleven a.m. is to submit to a kind of polite brawl, where fancy cameras are wielded at face-height and small, fashionable dogs scuttle underfoot. Children and buggies, perhaps inadvisably brought by optimistic parents, jostle through the throng of knees and, above it all, the pealing cries of the stallholders, many of them the latest in a long family line to shift cut flowers and potted plants on a Sunday morning. Since Shoreditch and Hoxton, the neighbourhoods that surround the market, transformed into hipster hangouts in the mid-Noughties and then into a kind of moneyed, gleaming tech start-up outpost of the neighbouring Square Mile a decade later, Columbia Road has become an established addition to a busy Sunday day out for tourists and Londoners alike. Nearby Brick Lane also transforms into a jostling market, the highly cynical and much-reviled Boxpark development offering all manner of brunch and drinking options. Bundles of long-stemmed flowers wrapped in brown paper and the swaying tips of a parlour palm are carried around on shoulders on pavements from Bank to Bethnal Green.

The flower market used to start at eight o'clock, end in the early afternoon. But the meandering, hungover idlers have pushed the timings back – the stalls don't finish setting up until nine-ish now and you can still pick up bargains at four. The demand for urban greenery – and Columbia Road remains the most accessible place in London to buy plants – has shaped decades-long tradition. The fashions defined by interior magazines and social media show up, a few months later, on the stalls: purple *Oxalis triangularis* (something that I have always grown by corm – the kind of tuberous, or bulb-like, beginning of several plants – because it was so difficult to get hold of) is now a regular; dozens of succulents in seedling pots scatter the street under a gazebo between the stalls of annuals and cut flowers; the indoor plants, which used to be an extra to the flowers, now comprise whole enterprises with fashionably pale pink-striped calatheas and doomed-to-death fiddle-leaf figs on sale for less than a round of drinks in one of the bars nearby. The market sells what the people want, and since the mid-2010s the millennial appetite for greenery, even that inside their rented city homes, has been painted on these pavements.

I was among them from the first summer I spent in London. Then, it was just to look. But I started to go to buy, to fulfil my desire for plants I could grow big enough to eat. That first early-twenties May was the beginning of a practice that I would come to hone as I became a gardener. I made a list and got down there early and alone. I did a trawl of the stalls (around thirty or so) only to return to those ones that looked good and best value to buy what I wanted. For many years Columbia Road was my main source of plants and taught me good lessons about buying them: to check that the soil is not sodden, the roots not drenched beyond survival, the surface not mildewy with wet; to aim for the plants heavy with bud rather than bloom and, if you are swayed by impulse, to make sure that it will thrive in whatever small space you are taking it home to. I would learn to get to the market regularly – increasingly early on a Sunday morning, escaping when the hordes turned up, getting to know my favourite stalls for annuals in trays and herbs in boxes and bulbs and hellebores, accordingly.

Maybe it is fitting for the young and dapper to have this impact on the sale of plants: Columbia Road Market started as an act of gentrification. Angela Georgina Burdett-Coutts, granddaughter (and heiress) of the banker to the aristocracy, built it as a covered food market in 1869 after buying what had become a notorious slum, crawling with murderers and gangs. Burdett-Coutts's market lasted a couple of decades before becoming a Sunday flower market – Sunday so that the local Jewish traders could work and the rival Covent Garden and Spitalfields sellers could ply their leftover stock, and flowers because of the legacy the Huguenots (bringers of auriculas, anemones, my ancestors) had left in the East End since their arrival 200 years earlier.

•

After the bluster and contrariness of April – alarmingly hot one weekend, only to fall cool and grey by Monday morning – it feels like spring finally settles and grows into itself with May. It is a time of final frosts, when gardeners start hardening out their crops, taking them out to test the cooler air and, after a few nights spent indoors, leaving them there to grow with the abundance of unfiltered natural daylight and rain. The sunshine becomes more steady, summer feels tangible instead of a distant, faintly unimaginable concept. Suddenly, as if participating in some mass overnight conspiracy, the trees green themselves, box-fresh leaves appearing from branches and filling the air with a sense of renewal so present you can smell it, of change, possibility and industry. Hawthorn, whether in compact bushes or unruly, ganglesome trees, dresses itself in pretty little white flowers. Wisteria drips pale purple globules across brick walls and houses fancy and council alike, sometimes clambering into other trees – a magic-trick disguise. Lilac throbs in hues of its name and perfume. The bits of lawn that are shared and seen by everyone – verges by roadsides and communal parkland – become ditzy with dandelion clocks, their sweet fluff remaining perfectly spherical until the next gusty day. And when the wind does come, it causes havoc. Plane and lime trees detonate with pollen that floats on the air, fragmented by traffic, the particles turn in tiny tornadoes before settling on shoulders and in gutters, accumulating in piles of white, foreign dust that leaves the prone in fits of sneezes. The day arrives early and with it a cacophony of chirruping and chanting: the dawn chorus. Even in London, even in the city, the collision of birdsong cuts through the rumble of the empty, desolate buses, the softly roaring skies. Nature is ripe and ready after months of patient preparation.

Life was surging into the city and my body and mind mirrored it. Perhaps May was the companion of September in my own inner calendar, a month of propulsion but also one steeped in reflection. To look back felt all the more necessary in the wake of such newness. I had always previously associated it with the onset of summer, of the nervous frenzy of exam season and the excitement of the warmer, longer days ahead. But this time around, in the wake of the break-up and all that had passed since, it became its most bittersweet. A whole year had passed since Josh and I had been truly happy for the last time. The anniversary of the break-up would bring its own untold significance. But somehow the annual marker of what I had believed to be turning a corner, a moment of golden possibility that actually proved to be hollow and false, was more difficult to face. The year before we went away for a weekend and it felt like a first date, infectious and thrilling and rich in desire. Outside London, beyond the repetition and rigour of our day jobs and free of the domesticity we had woven around one another, I was imbued with optimism. It was warm and sunny, and we went for walks holding hands. I had never considered that we would end, of course, but I knew that we had been through a rough patch and were, now, making it out the other side, that he and I were still capable of searing happiness.

And it was this that I mourned a year on. I spent weeks acknowledging a strange state of grief and nostalgia for that which I had previously known. The intoxicating blend of hope and denial that I had become so dependent on that I couldn't even entertain the reality of what was happening around me, the imminent destruction of the life I had mapped out so thoroughly. Since the break-up I had cycled through numbness, upset, anger, confusion and distance in wild degrees of severity, from dull aches to gnawing passions. But here was something different, a new kind of sadness and sympathy for the naive girl I had been, a condolence for her false, if undeterred, hope.

It made me wonder if I would have changed anything. If I would have gone back and shaken her, told her to examine things properly, to gimlet her eyes and remove her dependence and see that she could stand alone without the castles she had built around herself. If I would have tried to stop the denial and tune in to Josh's warnings.

I concluded that I wouldn't, that it had to have unfolded that way, that it had to rip me apart while I was still full of love and fierce protection for a man who was struggling to be with me, to share in the life that we had made. But, still. I was still sad for her, sad in the knowledge that that was the last of it, the last of blind, wild sentiment. That, from then on, many things would be tinged with the sense of impermanence, of knowing nothing that good could last for ever, of always being slightly prepared for it to be stolen away.

That impermanence, though, I had learned to recognise as a treasure. For all the residual melancholy, I thrust myself into May full of the dawning intention to claim that spring/summer as our own. One in which I would drink in the lovely things I had around me, to inhale the season as if it were nectar, sweet and vital and increasingly precious. I allowed myself ease and freedom; spent weeks ricocheting around Berlin – all swimming in rivers and lolling in parks, dusting linden pollen off black denim – and Barcelona with Heather and Jamie and a smattering of others, loving and laughing hard, sleeping less and being surprised by the new dawn arriving when we had outrun the night. The slews of admin and upset that had dogged April tempered for a while. I revelled in slapdash plans made possible by long evenings, fell in love with London's newly green and petal-laden streets again and again and again, watched a new crop of faint freckles scatter the bridge of my nose.

After years of treating the balcony as a military operation with the onset of spring – new annuals to fill the boxes with, seeds to be sown, bulbs to be lifted, tubs to be shifted – I sat back and realised that there was no need for seasonal fussing. The containers were growing into themselves. The winter hellebore flowers had turned pale green and firm, seed pods swelling with obscene productivity, new leaves turning up beneath them. The ferns, long dormant and scruffy, were offering new fiddleheads from the soil, furry and excitable as pups. The camellia's glory was fading, elegant white petals turning mushy and brown, landing face-forward on the grey concrete but leaving in their wake an abundance of shiny new leaves. Where before I would have wrenched out fading foliage, horrified by the seeming imperfection of it, I had become happy to let it be, realising that this was beautiful, space-filling stuff that was fit for purpose. I had become more content to be curious, letting things grow and seeing how beautifully they turned out, rather than constantly striving for control.

I got back from Barcelona and opened the door to find Matt in the kitchen. I had smelled dinner from down the corridor. I'd lent him my keys so he could water the seedlings and young, newly sun-drenched plants while I was away, and he'd made me something to come home to. The flat had become less of a home and more of a pleasant receptacle for life, no longer a cage or a place of box-ticking status but simply something I was lucky for, and soon would be out of. Another place where a chapter of my life had been. Here he was, here we were, occupying our own new domesticity. I hadn't failed at independence because he had come to look after me. I hadn't passed some test of millennial success because I'd spent the past few days partying in Europe. This was a quiet, unshowy thing, normal and elementary: that he was a man I loved who happened to love me, and that we made each other happy. After the tumult of our first few months together,

I was settling into place where he had been for far longer, of content existence. No room within it, often, for the big questions, much as I was prone to panic and propulsion, to fits of frustration at his steady spontaneity. I had long stopped asking for certainty or surety, but I was beginning to learn to exist in this state. To heed his oft-uttered advice of 'just be'.

The middle of the month was charged with downpours. April had been, with the exception of that one heaving weekend, a cool and dryish month. And so I greeted May's rain with the same pleasing familiarity as one might an old friend, one good enough to make themselves a cup of tea after letting themselves in. The patter of it tugged at my consciousness in the first light of an underwhelming dawn, and it returned again a night later. I would turn my head to feel it against my face, savouring the way it turned the air briefly clean. A few days on and the rain had arrived in earnest, rippling puddles by the time I made it to the window and lasted all day. I cycled to work in it, letting it drench my skin before peeling off sodden lycra to marvel at the semi-circle of charcoal roadspray etched on the back of my white T-shirt, and thus, I imagined, the skin that covered my spine. And the plants drank it in too, after so many weeks of chemical-imbued stuff straight from the tap. The seedlings had found their outside homes, now. The once-windowsill tomatoes were growing tall and confident, the sweet pea stems muscular, basil and parsley the first to be harvestable before a summer laden with greens.

The rain left freshness, wiped clean the balcony of city dust and pollen, sent the whole lot gurgling down the drain in the middle of my skyline plot. The midst of the downpour renders the air with nothing to smell but action. The wake of it, though, is delicious. It is amazing how strong that petrichor can be, even from plants crammed into concrete. How perfect it is in its nascence, how much promise it holds, that clean, earthy heft of stirred soil and new growth. How it transported me to the fertile shade of Jesmond Dene; to stepping out into that Hackney back garden in the quiet morning hours after the parties dwindled; to unlikely places where new life lay.

•

I spent different bits of my twenties in different parts of London. We rarely strayed into the west, although that first summer I often found myself in Clapham or even Balham, where some of my friends lived. The north-east reaches of Hackney pulled me out from Peckham for a few years, only for me to return to that corner once I tipped into the second half of that decade. It would frustrate and elate me, this city. How it tempted us in with its jewels, how it kept us out with glass walls and ceilings. But I rarely thought about leaving it. London is my most compelling lover; I am hooked on it, for all its mistreatment. Expensive and demanding and, if you know where to look, beautiful. In finding its green pockets, I felt I could tame it. So much life here thrived in spite of the hard-scrabble. I didn't need vast expanses or quiet rurality, I just needed pockets and green lungs big enough for me.

The river may be divisive but it is London's lifeblood, and so it has always been a constant in my history of the city. I'd cross the filth of the Thames on bike, on bus, very occasionally in car and on foot, the dark and churning murk of it sometimes reaching my nostrils. More rare and glittering was the flint of its grey waters on a sunny day, when the buildings that flanked it shone in the haze, filling me with gratitude to live here and not just be passing through.

Although it was riddled with tourists and out-of-towners, Southbank Centre, a gorgeous thud of mid-century brutalism, always lured in me and my friends. Along with the great box of the neighbouring National Theatre and the oft-forgotten BFI, Royal Festival Hall and the Hayward Gallery sat on the southern bank of the river. A brilliant eyesore, it stared defiantly at the overwrought fussiness of Somerset House, the Strand and Covent Garden beyond Hungerford and Waterloo Bridges. It was in this part that Josh and I first met, in a grotty pub opposite Waterloo Station. It was here that, cash-strapped, I would wander along the river, soaking up a city I couldn't really afford to be in. I have parked several generations of bikes in the plentiful racks underneath the criss-crossing concrete structures countless times; came to these buildings for literary parties and street food fairs, gallery openings and talks, gigs and plays and free events; watched the sun set from members' rooms I shouldn't have been in and dashed into the loos (the patina of their 1950s panelling still gleaming) after a long day out in town. Through them all I have chased luck and loss, wondered why I wasn't feeling more, why it couldn't be enough, wondered if it would ever get any better and, sometimes, felt that it was so good it could never be topped.

Often, I would stray up the yellow spiral staircase to the garden on the roof of Queen Elizabeth Hall. It takes a while to understand Southbank Centre as anything other than a labyrinth. It spans the distance between two bridges, there are multitudinous levels and seemingly magical sets of steps. It was a solid year before I realised that I had been taking a very long bus route from Peckham and walking from Blackfriars when I could have just hopped off another bus on Waterloo Bridge and arrived right there.

But I somehow always found the little expanse of lawn that felt like a secret. It was flanked by concrete – as everything was. Chunky, uncomfortable benches interrupted the green, but the view was riches: over and above the chaos in front of Festival Hall below, skimming over the music and the screaming children, the gaggles grabbing chairs to sit with expensive drinks, and beyond, down to Westminster and the glint of the London Eye carriages swaddled in clouds. To see those skies was to be birdlike and free in a city I often struggled to understand, to feel part of. Here was a small bit of ownership, a something that I had discovered.

At first, the gardens were just lawn; barely gardens at all, more a rare expanse of exposed and scrubby green in the heart of the grey city, with a bar inside a hut. The trees were too small to be noticeable. They had been opened in 2011, my first full year in London, and were treated by us, as by many, as a novelty drinking hole – there are few rooftops large enough in London to accommodate its masses, so craven and excitable in the presence of evening sunshine. But they grew as I did, welcoming in all manner of native species alongside ever-establishing olive trees. Chairs and patio tables, once clearly laid out to lure people to this budding space, became harder to spot. The large containers filled up with contented green sprawl, much of it edible. Rosemary flowered in tubs, flanking benches that stood against the concrete of the surrounding buildings. Lavender drank up the light. Umbellifers swayed above trailing nasturtium leaves. An arbour at the back of the garden grew heavy with leaf and blossom, a fertile artery in a sky full of hard lines and grey walls. On an often off-limits bridge between one building and another, in the narrow gap between paving slabs and chest-high walls, grew rare orchids. Here, surviving, in this most unlikely of places. On the quieter days, you could spot little business: bees humming against the thrum of the traffic, aphids and flies coalescing on new growth and the birds coming for them. All of it existing in boxes on man-made buildings.

All those springs and summers I had been coming here, silently drawn to this space, little understanding why. To drink Pimm's when it was warm and grab a patch of grass for an hour with a book. To fill the gap between work and evening plans, to have idle conversations and serious ones. At first I'd come for the benches at the front, the ones that gave you the view and sense of superiority. But with time I had been drawn further back, into the overgrowth and the insects, finding calm and measure among the green.

It had always been there, this need for nature. Instilled in me by my grandparents and theirs; surrounding my childhood so thoroughly I had become blind to it. City life subsumed it, as I wanted to learn the ways of pubs and bars and clubs and warehouse parties, of work and career and home and life and friends and lovers and long nights and early mornings and doing, in all these, the right thing. As I pushed at those, nature waited. Went through its processes, of germination, bud, flower and seed to scatter. Leaves appeared, grew, and fell. Winter held us cold, spring surprised us, summer went too quickly. And all the while it waited, while I was busy grabbing at what I thought life was. Then, when I thought I had got it, captured it all and kept it safe, nature made itself felt. I hankered for it, looked for it when all else was uncertain because I saw solidarity and comfort in its mysterious rhythms and unassuming ways. I wanted to translate its language, understand its patterns. In trying to, I realised I probably never would. Knew it would always surprise me, that every time something bloomed – even when fat in the bud, it would be a tiny marvel. It would catch me off-guard, like a warm room filled with basil breeze, or a smiling man on a balcony saying my name, and we too would begin a kind of clamouring love affair, relishing every moment that it would last for. And then it would change, become something different, and that would have its own kind of necessary charm, too.

They say a garden looks most beautiful a year after its custodian dies. That the plants, left to their own devices, break free of the confines placed against them. The strictures of pest control and pruning stop. Wilted flowers go un-deadheaded, seeds scatter the way they were intended to be. Surprises germinate, take root. Others burst free of their confines, no longer bound in place. Everything is a little wild and unruly, the measures and designs they were originally placed in become blurry. Path edges give in to weeds, flowers appear in the cracks. Blossom and fallen leaf gather under pots.

was the departure I had dreaded. Now that it was here it wasn't too bad at all.

The pots proved heaviest; we left them to the removal men, who hoiked the big tubs of ferns over their shoulders as if they were pillows. Jamie, his boyfriend and Matt, they picked up the boxes and packed the lift with them so swiftly I felt wildly unhelpful, opening doors and offering repeated, pathetic gratitude as my life as I knew it whisked out the door, leaving me caught in rippled air.

The flat I had found in April, the one in the woods, was to become mine after half a dozen near-misses. That was where the removal van was taking me, taking my half of the stuff. The others hadn't seen it, and when we got to the estate they ran up ahead of me to get a look. Jamie came down, told me it was perfect. I was still too scared to go back in, shy about making formal acquaintance with the place that would become my home. But then the removal truck left and it started to rain. The house plants and balcony tubs filled spaces in the car park and received the first natural drink of their lives before being moved in first. I walked in to find the balcony fully furnished. 'It looks like yours, now you've got all your plants here,' Matt said, landing a kiss on the side of my head. There was a small celebration: mismatched chairs popped out on the balcony for a picnic lunch and a bottle of Appletiser, passed around and swigged from the neck.

And then the boys left and I was there alone. I set to a couple of things, filling that time with activity: pulling down the curtains and lampshades to let in the light, nailing up a couple of hanging planters on the balcony and putting my grandfather's mirror on a hook on the wall. But then pause set in. There was so much to be done, to paint and strip and demolish and build. To buy and restore and covet. But nesting, that would take time and I was willing to give it; it was remarkable just to be here.

The summer had passed in the wrenches and stitches of transition, of scrabbling together money and documents, of calling solicitors and making diplomatic conversations. The balcony became smothered in leaves; I feasted on its produce, tomatoes and pea shoots, rocket and salad and herbs and edible flowers. I fled the city when I could, drank up my friendships as much as possible, tried to distract myself from the grinding and gritty process of physically removing myself from what was left of Josh and me. What we had been, and he, would stay in my head. Would visit my dreams and tinker with my daily thoughts. With time I came to realise that he would always occupy a bit of my brain and it was easier to accept that rather than push it away.

And now it was late September, and I knew enough to know that this new space was both a fresh start and a continuation of everything else that had happened. That I was still the same woman and a new one. I felt steadier, calmer, less fearful. I knew the balance I had on my own feet and trusted it, but also the deep reliance I had on sharing my mind and heart with others. Fewer expectations, happier to challenge them. My pride had shrunk, perhaps my dreams had too. And instead, I was better at taking in what I had and what was around me.

What was around me, what would become my balm and heartsease, were the trees beyond the balcony. Sure, it was a lovely view – the vast, shining expanse of city had been traded for something more light-absorbing, more gentle and bold – but it was more helpful as a constant reminder of the alchemy of my surroundings wherever I was. Of the weight of nature's silent science, of its indifference to our days that existed in every tree and plant and hedgerow in the city and suburb and beyond. I still knew little of what was to come. I knew I would try to take down the pine panelling in the kitchen and that the place would feel better with a lick of paint and an open window. I didn't know if the love that still held Matt and me in a giddy grasp would last or fade or shatter. I didn't know if my friendships would dampen or bloom, if my family would expand or shrink, if I would want for more in any of it. But I could acknowledge those uncertainties now. They would not break me.

What I did know, though, was that the leaves outside the windows of my new home would shrivel and yellow and fall. I knew that I would see beyond them, be allowed months to admire those strong skeletons left standing. That winter would blow in and be dark and cold and stultifying, and that spring would come and it would smell good. And I knew that I would watch it, and be surrounded by the seasons, and I would feel better.

ACKNOWLEDGEMENTS

BEFORE THIS BOOK, THERE WAS a newsletter, a kind of small and scratchy playground for my thoughts and ideas. The people who subscribed to it, and especially those who read it and replied, encouraged me to make something more of those words. Rachel Mills, who became my agent, was among them – and, importantly, was the first stranger to show such enthusiasm and vision for what *Rootbound* could become that I began to think it could be possible, too.

I'd like to thank everyone at Canongate for the care they have taken over this book, right from the start. I am particularly indebted to my editor Jo Dingley, who picked up this proposal and whose sharp mind helped me to become a far more thoughtful writer. Thanks, too, to Leila Cruickshank for her meticulous copy-editing. I have so appreciated the enthusiasm and support Lucy Zhon and Jamie Norman have given this book.

Rootbound was written alongside my day job, and that wouldn't have been possible without the quiet kindness of my colleagues. Thank you to Ross Jones, for endlessly picking up the slack, and to Serena Davies for making support a default. I remain grateful for Joanna Fortnam's generosity, without which little of this would have happened.

To those at the RHS Library who gave me space, time and the goodwill of locating hard-to-find books despite the fact I never carried a member's card.

On the matter of resource, the gardening community remain the most generous I know. Jack Wallington and Andrew O'Brien fielded so many plant fact-checking requests, even when they were working outside with soil on their hands.

To Charlotte Runcie, who made the most complicated things seem simple. To Amy Jones for her presence and sympathy over countless conversations, frustrations and writing sessions. To Anna Morris and Heather Welsh for maintaining the friendships that matter more than any book.

Thank you to my closest Vincents – Mum, Dad and Tom – for letting me get on with it and supporting me in the ways only they know how. And to Hannah Murphy, whose steadfast excitement has always meant so much.

And finally, to Matt Trueman, for his constant patience, motivation and belief.

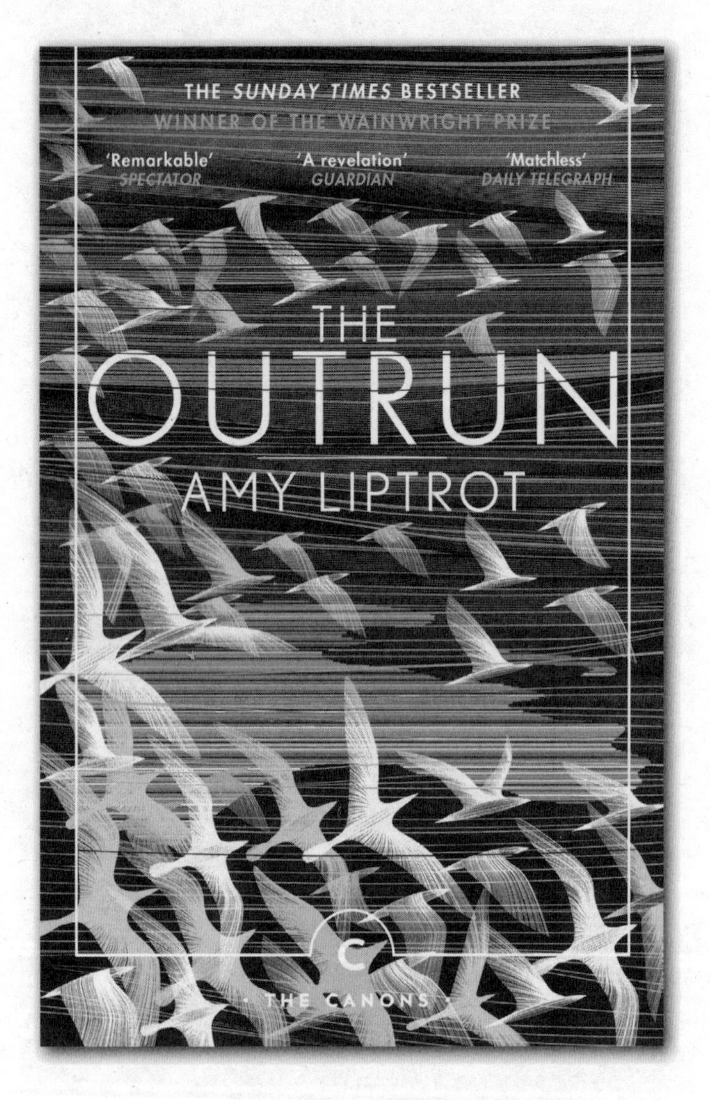

'A luminous, life-affirming book'

Olivia Laing

CANON GATE

'Astonishingly powerful' *Herald*

AS HEARD ON BBC RADIO 4

Charlotte Runcie

SALT ON YOUR TONGUE

Women and the Sea

'A joy' *Observer*

'Intoxicating' *The Times*

'Lyrical' *Economist*

'Bracing and poetic'
Observer

CANONGATE

'Luminously wise . . . Fierce and essential'
Helen Macdonald